Flood Recovery, Innovation and Response III

WIT*PRESS*

WIT Press publishes leading books in Science and Technology.
Visit our website for new and current list of titles.
www.witpress.com

WIT*eLibrary*

Home of the Transactions of the Wessex Institute.
Papers presented at FRIAR 2012 are archived in the WIT eLibrary
in volume 159 of WIT Transactions on Ecology and the Environment (ISSN: 1743-3541).
The WIT eLibrary provides the international scientific community with immediate and
permanent access to individual papers presented at WIT conferences.
Visit the WIT eLibrary at www.witpress.com.

THIRD INTERNATIONAL CONFERENCE ON FLOOD RECOVERY, INNOVATION AND RESPONSE

FRIAR 2012

CONFERENCE CHAIRMEN

D. Proverbs
University of the West of England, UK

S. Mambretti
Politecnico di Milano, Italy

C.A. Brebbia
Wessex Institute of Technology, UK

D. De Wrachien
State University of Milano, Italy

INTERNATIONAL SCIENTIFIC ADVISORY COMMITTEE

Organised by

Wessex Institute of Technology, UK

Politecnico di Milano, Italy

University of the West of England, UK

Sponsored by

WIT Transactions on Ecology and the Environment
International Journal of Sustainable Development and Planning
International Journal of Safety and Security Engineering

Supported by

EurAg Eng: European Society of Agricultural Engineers

CIGR: International Commission of Agricultural Engineering

ICID: International Commission of Irrigation and Drainage

WIT Transactions

Transactions Editor

Editorial Board

Flood Recovery, Innovation and Response III

Editors

D. Proverbs
University of the West of England, UK

S. Mambretti
Politecnico di Milano, Italy

C.A. Brebbia
Wessex Institute of Technology, UK

D. De Wrachien
State University of Milano, Italy

WIT PRESS Southampton, Boston

Editors:

D. Proverbs
University of the West of England, UK

C.A. Brebbia
Wessex Institute of Technology, UK

S. Mambretti
Politecnico di Milano, Italy

D. De Wrachien
State University of Milano, Italy

Published by

WIT Press
Ashurst Lodge, Ashurst, Southampton, SO40 7AA, UK
Tel: 44 (0) 238 029 3223; Fax: 44 (0) 238 029 2853
E-Mail: witpress@witpress.com
http://www.witpress.com

For USA, Canada and Mexico

Computational Mechanics Inc
25 Bridge Street, Billerica, MA 01821, USA
Tel: 978 667 5841; Fax: 978 667 7582
E-Mail: infousa@witpress.com
http://www.witpress.com

British Library Cataloguing-in-Publication Data

A Catalogue record for this book is available
from the British Library

ISBN: 978-1-84564-588-5
eISBN: 978-1-84564-589-2
ISSN: 1746-448X (print)
ISSN: 1743-3541 (on-line)

The texts of the papers in this volume were set individually by the authors or under their supervision.Only minor corrections to the text may have been carried out by the publisher.

Printed in Great Britain by Lightning Source, UK.

Preface

The present volume contains papers presented at the Third International Conference on Flood Recovery Innovation and Response (FRIAR) held in Dubrovnik. The conference started at the Institution of Civil Engineers in London in 2008 and was reconvened at the Lombardy Region in Milano in 2010.

The conference was jointly organised by the Wessex Institute of Technology, UK, Politecnico di Milano, Italy and the University of the West of England, Bristol, UK; sponsored by WIT Transactions on Ecology and the Environment, and the International Journals of Sustainable Development and Planning, and Safety and Security Engineering; and supported by the European Society of Agricultural Engineers (EurAg Eng), the International Commission of Agricultural Engineering (CIGR) and the International Commission of Irrigation and Drainage (ICID).

The occurrence of floods is the most frequent among all natural disasters globally. In 2010 alone, 178 million people were affected by floods. Losses in exceptional years such as 1998 and 2010 can exceed $40 billion. In 2010-11, destructive floods occurred in Pakistan, Queensland, Australia, South Africa, Sri Lanka, the Philippines, Brazil, Japan and the US. On average over the past three decades, more than one hundred million people each year have been affected by floods. The number of people affected has grown since 1950 from around four million a year to the present level, which represents more than 1% of the global population.

The impact of these events encompass tragic loss of life, damage to built and natural environments and massive disruption to the lives of affected populations in the short term. In the longer term the recovery and post recovery phases can also cause distress, disruption, health problems and financial hardship lasting many years. It has been calculated that over a period of 30-years floods killed in excess of 200,000 people directly and may have indirectly contributed to millions of illness related deaths.

Flood risk is seen to be increasing worldwide as a consequence of climate change, rapid urbanisation and warming seas leading to sea level rise. The close link between rivers and coasts and human prosperity has led to human settlement patterns that leave many major cities and other urban areas exposed to regular or occasional flooding. Recent population growth has meant that urban areas have expanded massively. Urban development has frequently proceeded without consideration of flood risk. Between 1950 and 2000 the urban population almost quadrupled to just less than three billion. Over half the world population is now resident in urban settlements; by 2025 some estimate that two thirds of the population will live in cities and towns, i.e. more than four billion individuals, 90% of these in developing countries. This has resulted in an increase in the population and associated buildings and infrastructure at risk from flooding. Increased wealth has also led to an increase in the value of property exposed to flood hazard. As a result, the insured losses from flood events continue to grow and will do so for the foreseeable future even without the predicted increase in flood hazard.

While meteorological events cannot be changed- (at least in the short term) - the manifestation of a flood event as a result of weather extremes is to some extent controllable. The design and operation of the urban environment can make a large contribution to mitigating the effect of weather, indeed it is one of its prime purposes. The driver for increasing recent interest in adaptation of the built environment has been the rise of extreme events which have multiple causes related to increased flood vulnerability and exposure.

This conference provides a forum for researchers, academics and practitioners actively involved in improving and interchanging knowledge and expertise in a wide range of technical, social and economic issues related to the broad themes of flood recovery, innovation and response.

The scientific topics presented at the conference include: Flood risk management; Flood risk vulnerability; Emergency preparedness and response; Flood forecasting; Flood case studies; Responses to reduce vulnerability to flooding, amongst others.

The Editors are very grateful to the Authors for their excellent contributions, as well as to the members of the International Scientific Advisory Committee for their help in reviewing the papers. The quality of the material makes the volume a most valuable and up-to-date tool for professionals, scientists, and managers to appreciate the state-of-the-art in developing a fully integrated approach to flood risk management.

The Editors,
Dubrovnik, 2012

Contents

Section 3: Emergency preparedness and response

Section 4: Flood forecasting

Section 5: Flood case studies

Section 6: Responses to reduce vulnerability to flooding (Special session organised by J. Lamond)

Section 1
Flood risk management

Ontology for flood management: a proposal

D. De Wrachien[1], J. Garrido[2], S. Mambretti[3] & I. Requena[2]
[1]*Department. of Agricultural Engineering,*
State University of Milan, Italy
[2]*Department of Computer Science and Artificial Intelligence,*
University of Granada, Spain
[3]*DIIAR, Politecnico di Milano, Italy*

Abstract

Floods are very complex phenomena involving a large number of players. In this context not only the hydrologic and hydraulic studies are required, but also those related to vulnerability assessment, town and emergency planning as well as all the matters related to the economic issues: damage, insurance and the psychological diseases which can arise after flood events.

Thus, people with widely different roles, backgrounds, and profiles are in charge of each phase of a large number of complex procedures. Not surprisingly, communication problems often arise because different names may be used to refer to the same concepts. Moreover, relevant information can vary, depending on the user.

Effective communication is essential since the results produced by a person in a certain phase of the process can be the input that another person needs in the next phase, i.e. raw data versus elaborated data, forecasting versus impact evaluation, level of detail, hazard assessment, vulnerability assessment, economic evaluation, environmental assessment, etc.

Ontologies, which can be defined as an explicit representation of a conceptualization, are tools that can be used to solve such very complex and interdisciplinary problems because they specify a conceptual framework or terminology.

The paper introduces an ontology for floods. This ontology is developed using the existing Environmental Impact Assessment (EIA) ontology and it is represented with OWL-DL (Web Ontology Language – Description Logics).

Among the different causes for flood, only those related to river overflows are taken into account in the ontology so far. It is shown as ontologies are useful

WIT Transactions on Ecology and The Environment, Vol 159, © 2012 WIT Press
www.witpress.com, ISSN 1743-3541 (on-line)
doi:10.2495/FRIAR120011

because they represent the most meaningful knowledge associated with a problem. Moreover, they are also useful because they permit reasoning and inference processes. Inference mechanisms make recommendation procedures possible.
Keywords: hazards, floods, knowledge representation, brief ontology.

1 Introduction

The terms "flood" and "flooding" are often used in different ways. According to International Commission on Irrigation and Drainage (ICID) [1] flood is "a temporary condition of surface water (river, lake, sea) in which the water level and/or discharge exceed a certain value, thereby escaping their normal confines". Flooding is defined "as the overflowing or failing of the normal confines of a river, stream, lake, canal, sea or accumulation of water as a result of heavy precipitation where drains are lacking or their discharge capacity is exceeded".

Although flooding is a serious hazard in humid regions, it can be devastating also in semiarid regions, where high rates of runoff following storms produce widespread flood damage down valley. Recurring floods are also typical in coastal and estuarine zones [2].

To cope with these hazards, it is imperative that human society adopts an effective flood hazard management approach which has to be in harmonious coexistence with floods. In practical terms, the chance of flooding can never be eliminated entirely. However, the consequences of flooding can be mitigated by appropriate behaviour and actions. To be effective, the hazard approach must be embodied in the broader context of integrated catchment planning, and flood must be regarded as one of the many issues involved in the appropriate management of a catchment [3].

People with widely different roles, backgrounds, and profiles are in charge of these different issues and therefore communication problems may arise because of the different definitions apply to the same concepts. Moreover, relevant information can vary, depending on the user, and the results produced by a person in a certain phase of the process are the input that another person needs in the next phase.

Ontologies are a tool that can be used to solve such problems because they specify a conceptual framework or terminology.

Although there are a wide variety of definitions of *ontology*, many of them are too focused on a specific application or require a background in logic. However, according to Gruber [4] "a body of formally represented knowledge is based on a conceptualization: the objects, concepts, and other entities that are assumed to exist in some area of interest and the relationships that hold among them". This Author goes on to define ontology as an explicit representation of a conceptualization. This definition was considered sufficient for the purposes of our study.

Ontologies also provide explicit definitions and restrictions for concepts, i.e. Knowledge Representation (KR). They can be used to describe a context or the domain of a context [5]. The Environmental Assessment Ontology [6] provides a valid conceptual framework for modelling flood-related knowledge. The paper shows the inclusion of a semantic model for flood management in the EIA ontology and the generation of a brief ontology in order to simplify and optimize the use of the ontology for flood information. It describes the semantic modelling of floods and flood-related concepts [7, 8].

2 Flood management

Risk is an integral part of social and economic processes and is often increased by human interference with natural hydro-meteorological phenomena. The struggle against extreme events like floods and droughts is as old as mankind. In the last decades, new challenges are likely to influence risk management measures and policies:

- increasing world population and economic growth lead to a more intense use of water and land resources;
- there is a rising awareness of the need of integrated flood risk management, considering the river basin as the basic planning unit;
- due to the relentless urbanisation process, at worldwide level, hazards are increasingly transforming into disasters putting development at risk;
- in flood prone areas there may be the problem of land subsidence. This problem makes such areas increasingly vulnerable;
- climate change is likely to impact climate variability, making extreme events more severe and more frequent even if, compared to the items listed above, impacts of climate change are of limited importance;
- there is a rising concern that damages resulting from water related disasters are growing disproportionately worldwide.

To cope with these challenges involves taking decisions and actions about appropriate levels of risks. These decisions and actions may be divided into risk analysis procedures and risk management cycle.

In recent years risk analysis has emerged as one of the most appropriate methods nowadays available to assess natural hazards, like floods and other water related disasters. This methodology proved to be comparatively reliable in determining the hazard potential and the related probability of occurrence of defined extreme events, but this process requires expert knowledge to identify the potential risks and then to estimate the likelihood and the impacts of these risks. Risk likelihood and risk impact comprise uncertainty. In this context, the last two decades have observed an amazing progress in the use of probability theory to assess uncertainty. However, floods are natural hazards, which occur periodically and episodically and cannot be prevented.

Therefore, the responsible authorities in most European countries developed methods of integrated risk management, which follow, mainly, the approach to express risk as a product of hazard and values at risks. To this end, it is worth noting that, whilst in the past the concept of risk was primarily intended as a measure of the probability of a system's failure, it has got, nowadays a more complex meaning. The risk has to be considered as a combination of both the probability and the magnitude of the consequences of a system's failure, and so, as the mathematical expectation of the consequence, taking into account of all significant hazards and all significant mechanisms of failure [9].

With regards to the risk management, according to the International Strategy for Disaster Reduction (ISDR) [10], its cycle comprises "the systematic process, administrative decisions, organisation, operational skills and abilities to implement policies, strategies and coping capacities of the society and communities to lessen the impact of natural hazards and related environmental and technological disasters". This covers all forms of activities, including structural and non – structural measures to avoid or to limit adverse effects of hazards.

On the whole, the risk management approach consists of systemic actions set up in a cycle of preparedness, response and recovery that should form part of any integrated flood risk management.

Preparedness consists of preventive and precautionary measures to prepare for an event before it occurs. It aims at minimizing the effect of development activities on accentuating the magnitude of hazards, reducing the exposure to natural hazards and minimising the socio – economic vulnerability of people and material assets exposed to these hazards. Response consists of measures that limit the effects of exposure to a hazard and its duration. It mainly focuses on alerting potentially affected people, rescuing victims and providing assistance in case of need. The recovery phase aims at enabling the economic and social activities to return to normal with a minimum delay. It also involves the analysis of the disaster in order to learn lessons and integrate corrective measures into prevention and preparedness plans. To this end, it is important to underline that the effectiveness of the risk management cycle in reducing risks and damages depends, also, on the political will to apply the risk management principles in developing planning, the existence of well defined institutional responsibilities and on a democratic process of consultation and social control with effective governance.

The challenge before the international community is to support these activities, particularly in least developed countries, where resources are limited, by means of actions aiming at:

- informing policy makers and the public of the trends in water – related risks and policy options to mitigate those risks;
- introducing long–term land use planning, taking into account the concerned aspects of flood risk management (including risk assessments) and adaptive management to reduce vulnerability to risks, that may be increase in time, due to the processes as described before;
- raising awareness of water–related hazards and improving the capacity of communities to respond effectively;
- developing conventional and state–of–the–art technologies and monitoring systems tailored to local conditions for water–hazard alerts;
- fostering specific capacity development programmes for water managers.

Finally, it is known as many players are involved in flood management, each of them having different backgrounds, skills and expectations. There is a need for a planning framework able to explain all the relations between the players and the role of each of them.

3 Ontologies

Depending on the application and context where the ontologies are used, they provide a mechanism to share knowledge by using a common vocabulary; they allow semantic labelling, exchanging and reusing knowledge; they can be also used to establish a communication protocol. Its formal background makes possible the semantic, logic and formal descriptions without additional tools. Its use in genetics has spread to fields such as knowledge engineering, artificial intelligence and computer science, language processing, knowledge representation, cooperative information systems, ecommerce, bioinformatics, database design and integration, smart integration of information, information retrieval and knowledge management.

Roughly speaking, an ontology consists of properties, classes, instances and operators:

- Classes correspond to concepts or elements in the real world.
- Properties are relationships whose domain and range are defined in terms of concepts.
- Instances are individuals of classes.
- Operators represent the allowed symbols which provide the set of operations or mechanisms for the concept definitions.

The concepts can be formally defined by axioms using the operators, properties, instances and other concepts. For example, a concept can be defined as the union of other concepts.

The concepts' definitions are supported by the Description Logic (DL), which are a family of knowledge representation languages. The set of operators and its use implies different levels of complexity and efficiency for the description logic. In general, lower expressiveness of the language implies higher efficiency but poorer definitions whereas higher expressiveness brings more flexibility in the definitions but decidability problems in the worse scenario i.e. it is not guaranteed to find an answer for a question in a finite amount of time.

Although there are several languages to represent ontologies, OWL (Web Ontology Language) is the current standard according to the W3C (World Wide Web Consortium). The W3C is an international community that develops open standards to ensure the long-term growth of the Web. The set of operators in OWL consists in union, intersection, negation, one-of, cardinality, universal and existential operator.

Beyond the concept definition, there is a structural relationship which allows representing the concepts as taxonomy. This relationship is referred as IsA and provides a hierarchical representation of the concepts. Therefore, a concept will always have a parent and if two concepts are children of the same parent then they are called siblings. Additionally, if a concept (X) is the parent of other concept (Z) then the concept Z is generalized or subsumed by the concept X (commonly said that Z is a (IsA) X).

An ontology which only uses the IsA relationship is a taxonomy of concepts. However, we take advantage of ontologies by describing formal definitions because it allows implementing semantics and using reasoners. The reasoners are applications that check the consistency in the ontology i.e. there are no contradictions. They also make possible the inference mechanisms to discover new knowledge or launch queries over the ontology.

Let us suppose an ontology where the concepts Country and City are defined in form of classes. The individuals of these classes will be the instances of the concepts i.e. the current countries and their cities. If there is a property called hasCapital whose semantics implies that a country has a capital then the definition of the class Country could include a restriction to force that a Country must have a Capital. The example bellow shows how the countries Italy and France are related to their capitals.

- Class: City
- Class: Country (∃ hasCapital, …)
- Instance of City: Rome
- Instance of City: Paris
- Instance of City: Barcelona
- Instance of Country: Italy (hasCapital Rome, …)
- Instance of Country: France (hasCapital Paris, …)
- Property: hasCapital (domain: Country, range: City)

Considering that all the cities are not capital, a concept Capital could be defined as anything that is capital of a country by defining the property isCapitalOf (domain: City, range: Country) as inverse of hasCapital: Capital (∃ isCapitalOf). If we do so, the instances Rome and Paris will be additionally inferred as individuals of the concept Capital.

There are many tutorials and literature about ontologies, for example, the tutorial for Protègè (an ontology editor [11]). Additionally, some books gather a list of applications based on ontologies and show how they are used [12]. Finally, as the complete ontology may be slow and difficult to deal with, because it includes a lot of knowledge that is not related directly to floods, brief ontologies may be built [13].

4 Building an ontology for flood management

An ontology can be used to model floods because of the close interrelation between the elements involved. Other approaches, such as databases, are not satisfactory since they only store data. However, our objective was to provide the representation of an entire system. This necessarily includes a specification of the connections between its components so that the model can perform reasoning tasks.

Although there is no standard for ontology development, there are principles and recommendations that can be followed. One of the most well-known set of guidelines is Methontology [14], which recommends reusing ontologies to the extent possible. Besides saving time and resources, this makes integration between applications considerably easier.

In particular, the methodology we have implemented is organized as follows:

1. Identification of the aim and the scope of the ontology: it can impose requirements in the construction phase.
2. The ontology construction process:
    a. Getting knowledge: identification of concepts and relationships that should be in the ontology and design of a strategy for this task.
    b. Encoding phase: defining classes, individuals and properties in OWL.
    c. Integration with other ontologies in order to find an ontology with similar domain that can be reused or that can be specialized.
    d. Inference in order to check consistency and correctness.
    e. Creation of brief ontologies to extract relevant portions of the ontology for more specific domains.
3. Evaluation (by experts) and maintenance.

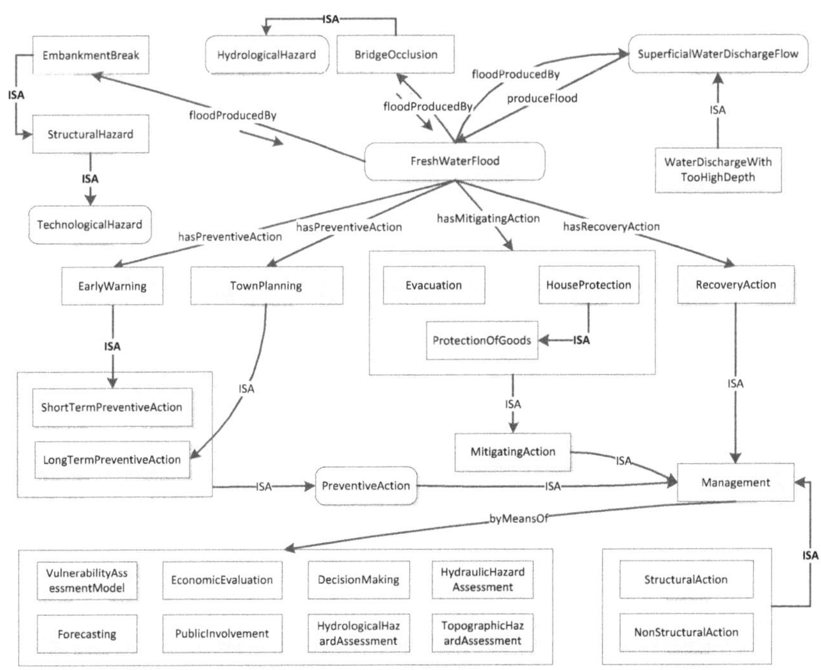

Figure 1: Conceptual schematization flood.

According to the previous methodology, the EIA ontology was chosen as a starting point for flood modelling because many of the concepts needed for such a representation are already in the EIA ontology. Some are directly included, whereas others are generalized in the representations of other concepts. In such cases, the concept can be included by adding more details to the existing taxonomy or by establishing new relationships to describe food-related knowledge. Alternately, new concepts can be added to the first level of the hierarchy.

In figure 1 the schematization of flood is shown. In the figure, nodes (rectangles) represent concepts, and they are linked by relations.

Flood is defined in the European Directive on assessment and management of flood risks as the temporary covering by water of land not normally covered by water. This general definition refers to all type of floods: river floods, flash floods, urban floods, and floods from the sea in coastal areas. Although all of these flood types share certain characteristics, this paper focuses on freshwater floods and particularly on river floods. For this reason, we have defined flood as a body of water that overflows its usual boundaries onto a land area with other land uses. This produces negative impacts caused by water velocity, depth, persistence, or any combination of the three.

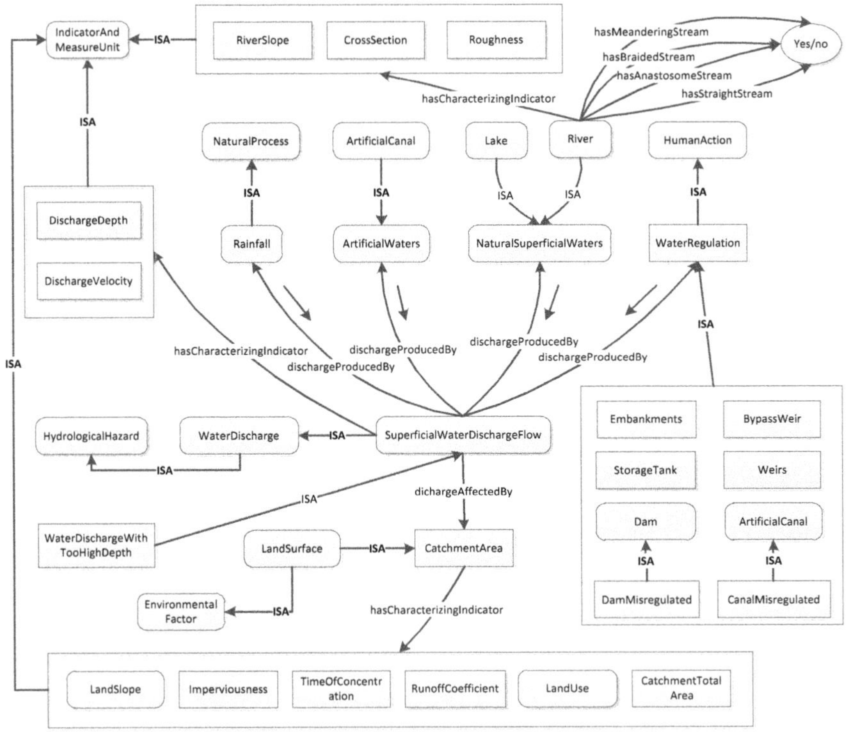

Figure 2: Conceptual schematization of superficial water discharge.

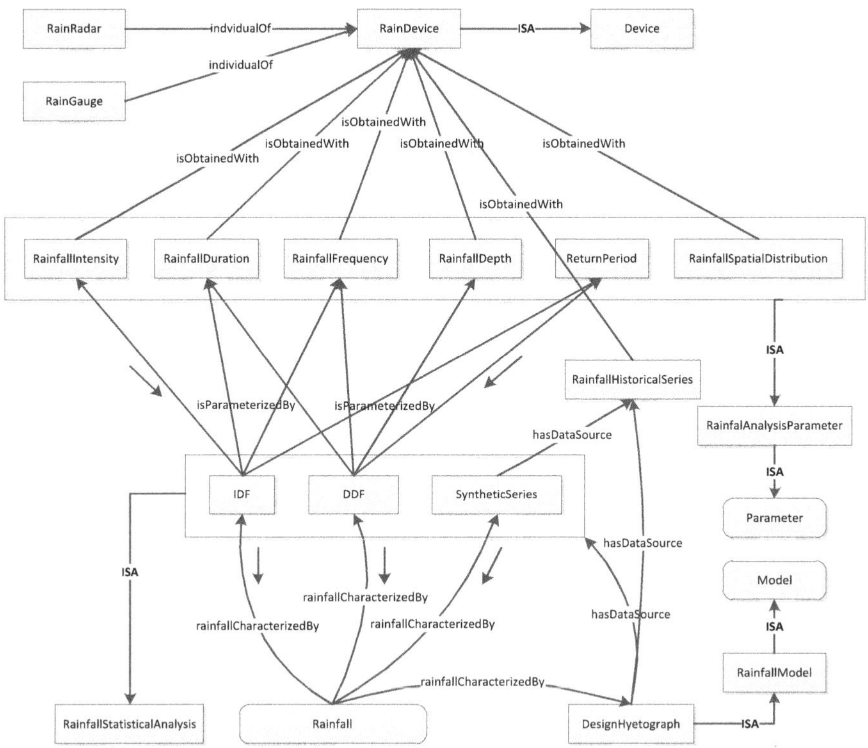

Figure 3: Conceptual schematization of rainfall.

Each of the concepts can be (and should be) expanded, as each of them is quite complex and includes a number of subconcepts and relations. As example, in figure 2 is reported a preliminary conceptual schematization of superficial water discharge flow, which appears in figure 1 as one of the causes of floods. Again, many events can cause a discharge in a river or in a channel; one of these is the rainfall, which is a concept that can be deepened and the result is shown in figure 3. Once this job is carried out for all concepts, and the relations between concepts are defined as well, a complete representation of the real world will be available.

5 Concluding remarks

Floods are complex phenomena involving a large number of players and therefore many studies are required, such as hydrologic and hydraulic, vulnerability assessment, town and emergency planning as well as all the matters related to the economic issues. Thus, people with widely different roles, backgrounds, and profiles are in charge of each phase of a large number of complex procedures and this can produce communication problems, especially because the concept of "relevant information" varies depending on the user.

Moreover, the results produced by a person in a certain phase of the process can be the input that another person needs in the next phase.

Ontologies, which can be defined as an explicit representation of a conceptualization, are tools that can be used to solve such very complex and interdisciplinary problems because they specify a conceptual framework or terminology.

The paper introduces an ontology for floods. This ontology is developed using the existing Environmental Impact Assessment (EIA) ontology and it is represented with OWL-DL (Web Ontology Language – Description Logics). In the paper, first the conceptual framework for floods is shown. Here it is shown the reasons why floods may happen and how the different actions to be taken to reduce their consequences are related. Among the different causes for flood, only those related to river overflows are taken into account in the ontology so far. Then, as the river overflows are considered for floods, this concept has been expanded showing all subconcepts and properties related to it.

Finally, as an example, one of the concepts shown in the conceptual schematization of superficial water discharge (namely: rainfall) is expanded again in order to show the level of detail and deepness that can be reached in describing concepts. It is shown as ontologies are useful because they represent the most meaningful knowledge associated with a problem. Moreover, they are also useful because they permit reasoning and inference processes. Inference mechanisms make recommendation procedures possible.

References

[1] Schultz B. 2001: Irrigation, drainage and flood protection in a rapidly changing world. *Irrigation and Drainage* 50(4), pp. 261-277.

[2] Varnes D.J., and IAEG Commission on Landslides and other Mass Movements and landslide hazard zonation 1984: *A review of principles and practice*, 63 pp, UNESCO Press, Paris.

[3] Meadcroft I.C., Reeve D.E., Allop N.W.H., Diment R.P., Cross J. 1996: Development of new risk assessment procedures for coastal structures. *Advances in Coastal Structures and Breakwaters*, Thelford T. (Ed.), London, pp. 6-25.

[4] Gruber, T.R. 1993: *Toward Principles for the Design of Ontologies Used for Knowledge Sharing* Technical Report KSL 93-04 Knowledge Systems Laboratory Stanford University.

[5] Dey, A. & Abowd, G. 2000: Towards a better understanding of context and context-awareness. *Proceedings of the Workshop on the What, Who, Where, When and How of Context-Awareness* (CHI), The Hague, Netherlands.

[6] Garrido, J. & Requena, I. 2011: Proposal of ontology for environmental impact assessment. An application with knowledge mobilization. *Expert System with Applications*, 38 pp. 2462-2472.

[7] Garrido J., Requena I., Mambretti S. 2011: Building brief ontologies. A case study for floods management. *Proceedings of the International*

Conference on Knowledge Engineering and Ontology Development, KEOD 2011, Paris, France, 26-29.

[8] Garrido J., Requena I., Mambretti S.: Semantic model for flood management Accepted for publication on the *Journal of Hydroinformatics* on 12th December 2011.

[9] Mambretti S., De Wrachien D., Sole A. 2008: Flood – risk assessment and hazard mitigation measures: Case studies and lessons learnt in Italy *Proceedings of the International Conference on Flood Recovery Innovation and Response*, Institute of Civil Engineers, London, United Kingdom.

[10] International Strategy for Disaster Reduction (ISDR) 2004: *Terminology: Basin terms of disaster reduction.* Updated version.

[11] Noy, N. F., McGuinness, D. L. 2005: Ontology Development 101: A Guide to Creating Your First Ontology. Stanford University, http://protege.stanford.edu

[12] Staab, S., Studer, R. 2004: *Handbook on Ontologies*. Springer

[13] Garrido J., Requena I. 2012: Towards summarising knowledge: Brief ontologies. *Expert Systems with Application*, V 39, 3, pp 3213-3222.

[14] Fernández, M., Gómez, A., Pazos, J. & Pazos, A. 1999: Ontology of tasks and methods. *IEEE Intelligent Systems and Their Applications*, 14(1), pp. 37-46.

Cooperation between non-governmental organizations and the State in the matter of flood risk management in the Czech Republic

J. Dostál
Department of Public Economics, Faculty of Economics and Administration, Masaryk University, Czech Republic

Abstract

The purpose of this paper is to identify the current form of cooperation between non-governmental organizations (NGOs) and the State in flood risk management in the Czech Republic. The institutional framework developed for the broad cooperation between the State and the nonprofit sector was applied to flood risk management as a means of examining this area. The institutional framework for NGOs–State cooperation in the case of Czech flood risk government has the following elements: the European Union, Central Government, Ministries, Regional Authorities, Municipal Authorities, Emergency Services of Integrated Rescue System, Qua-volunteers, Employment Agencies, and State Companies. The methodology was qualitative and revolved around a literature review and managed interviews with six prominent NGOs participating in Czech flood risk management. Further data was collected from Czech professionals' conferences and during the author's membership in the working group of the Ministry of Interior of the Czech Republic. The conclusion of the paper is as follows: The overview of the cooperation could be helpful for other post-socialistic countries which face a similar situation as did the Czech Republic Moreover, an overview of all forms of NGOs–State cooperation in the case of Czech flood risk management could be helpful for Czech NGOs and the Czech State where cooperation is variable at all levels of governance across the country.
Keywords: flood, Czech Republic, flood risk management, non-governmental organizations, state, cooperation, institutional framework.

www.witpress.com, ISSN 1743-3541 (on-line)
doi:10.2495/FRIAR120021

1 Introduction

Palubinskas [1] notes that the implementation of communism in Central and Eastern Europe resulted in the virtual elimination of NGOs. After the establishment of democracy, Czechoslovakia (and later the Czech Republic) faced the challenge of how to involve NGOs in flood risk management. This became a particular focus in 1997 after the first big flood event in the Czech Republic following the demise of communist rule.

Czechoslovakia was not an exception. Many Eastern Bloc countries also faced a virtually eliminated NGO sector, a situation that could be considered point zero of NGOs' participation in flood risk management. The Czech Republic faced two challenges: first to establish civil society and secondly to manage a new type of flood, which affected the country. These two challenges merge into one in the case of participation of NGOs into Czech flood risk management.

An examination of the academic literature reveals a distinct gap on this topic. Studies by Frič *et al.* [2], Brhlíková [3], Hyánek and Pospíšil [4] and the others focus on non-profit sector in the Czech Republic in general, while Rektořík [5], Antušák and Kopecký [6] focus on emergency management in the Czech Republic in general. Kavan [7] writes about civil protection in the Czech Republic and also partly about cooperation between NGOs and Czech state in case of emergency management especially about the cooperation between NGOs and Fire Rescue Service of the Czech Republic. But he highlights one significant fact; both NGOs and Czech state need each other and are dependent on each other today.

In case of Czech flood risk management, studies focusing on NGOs were almost nonexistent. One exception is Kumar [8], who wrote about the flood in 2002 and focused particularly on coordination of humanitarian help of NGOs in the case of flood. But that is only a small piece of NGO-state cooperation in flood risk management. Moreover situation in Czech flood risk management and NGOs participation and cooperation with state in year 2011 or 2012 appears to be very different than in 2002.

During the review of the literature, literature presenting complex reviews of cooperation between NGOs and regional and municipal governments in case of flood risk management was not found. The most valuable literary source was the research report from Rektořík *et al.* [9], written by a collective of scholars including the author. This report has several parts; one of them was regarding NGOs and their cooperation with local authorities.

This paper's purpose is to identify a current form of NGOs–State cooperation in terms of Czech flood risk management. This was achieved through application of the theoretical approach with a general institutional framework to the current situation of Czech flood risk management.

2 Material and methods

Methodology was qualitative and revolved around a literature review, managed interviews and information gathered during sessions of relevant working group of Ministry of Interior of the Czech Republic.

Primary data was collected through managed interviews with six prominent NGOs participating in Czech flood risk management. The interviews took place during the first half of year 2011. The Snowball method was used for choosing the NGOs, starting with the most respected ones, and then moving on to others along the chain. The circle was closed at n=6. The list of the chosen NGOs can be found in table 1. Although there are many other NGOs participating in the Czech flood risk management process, the NGOs selected appear to be the most prominent participants.

Table 1: NGOs which take part in the research.

NGO	Position of the NGO representatives
ADRA	Coordinator of the humanitarian help in the Czech Republic; Economic director
Czech Red Cross	President
People in Need	Coordinator and head of flood programs.
Diaconia	Person, responsible for humanitarian aid.
Hand for Help	Managing director
Caritas Czech Republic	Head of humanitarian department – Archdiocesan Caritas Olomouc

After choosing the NGOs for managed interviews, it was necessary to identify relevant positions and the persons who occupy them to be interviewed. Because of need to make the findings relevant, of the findings, simple logic was used: as high relevant position as possible. The result of these efforts was that managed interviews were achieved with two heads of NGOs, one economic director and in four cases, with persons responsible for humanitarian aid in the Czech Republic. Positions of the NGOs representatives, which were asked during the managed interviews, are available in the following table.

All the NGOs were extremely willing to cooperate to participate in this research, including the CEOs. These interviews with NGOs were a central part of the research because they represent the main players in this field.

After the first part of research in June 2011, the author was invited to be a member of a working group of Ministry of Interior of the Czech Republic called Volunteering and Emergency and Crisis situation. During the participation and membership of the working party additional information was able to be gathered.

3 Results

Bullain and Toftisova's [10] institutional framework, in its original general terms was used. Their framework has eight levels of NGO cooperation with government: Parliament, Central Government, Ministries, Agencies and authorities, Councils or joint committees, Quangos, Specific Bodies and Local form of cooperation. In this paper you can see the application of this framework on current situation in Czech flood risk management. Because of the quite specific character of this part of non-profit sector, the general framework was modified.

The new institutional framework for NGOs–state cooperation in the case of Czech flood risk management has the following elements: European Union, Central Government, Ministries, Regional Authorities, Municipal Authorities, Emergency Services of Integrated Rescue System, Qua-volunteers, Employment Agencies and State or Half-state Companies.

3.1 European Union

A special case of cooperation of Czech NGOs with a central form of government is the cooperation with the European Union. NGOs could use cooperation with the European Union as a strategy for gaining financial recourses for Czech flood risk management work.

The truth is that the NGOs involved in Czech flood risk management have had some experiences with projects supported by Structural Funds, but in most cases these projects are from areas not related to flood risk management activities. For example, ADRA has a project for assisting the victims of domestic violence and working with children with behavioral problems. The only project supported by Structural Funds related to flood risk management, are projects of Czech Red Cross for disaster preparedness.

3.2 Central government

The central government in this context means the central government of the Czech Republic and its councils.

3.2.1 Central taskforce

The more important council of the central government is the National Security Council. The working body of the National Security Council which is important for cooperating with NGOs in case of Czech flood risk management is called Central Taskforce. It was found out that Czech Red Cross had already taken part in this session. This is an example of another form of cooperation between NGOs and the Central Government which impacts on Czech flood risk management.

3.2.2 Government council for non-governmental non-profit organisations

The second relevant Government Council is the Council for Non-Governmental /Non-Profit Organisations. The council has several committees that include the Committee for Regions whose role it is to encourage the NGOs–Regional government's relations and The Committee for Legislation and Finance whose role it is to regulation of the non-profit sector [11]. Previously three laws were prominent in relation to NGO participation in flood risk management: the law on public benefit, law on financial collection and civil code. One of the present members of this Council is the director of Caritas Czech Republic which is one of the primary NGOs participating in Czech flood risk management [12].

3.3 Ministries

The cooperation between NGOs and ministries has several forms. In general, they include the accreditation service and provide grants and organize the working groups of the Ministry.

3.3.1 Accreditation service and providing grants

The Ministry of the Interior operates an accreditation service for NGOs providing voluntary assistance. The accreditation system is closely connected to the provision of grants to NGOs. The accreditation and grant system is used by various kinds of services, not only for volunteerism as a part of flood risk management. There are two commissions for that purpose: Accreditation Commission and the Grant Commission.

Both of them have 9 members and some of the members are the same. The members of the commission are also representatives of other ministries (Foreign Affairs; Labour and Social Affairs; Finance; Culture; Education, Youth and Sport; Environment) [13]. Therefore it can be said that by meeting the rules of these commissions and communication with them, NGOs participating in Czech flood risk management are also cooperating with these other ministries, through its representatives.

In 2011 it was found that several grants from Czech ministries relating to Czech flood risk management to NGOs came from sources of ministries. The Ministry of the Interior, for example, provided a financial contribution to ADRA for crisis and emergency plans. The Ministry of Education, Youth and Sport also gave financial assistance to ADRA for activities related to the organization of European Year of Volunteering in the Czech Republic – 50 000 CZK). Some NGOs receive grants from the General Directorate of the Fire Rescue Service of the Czech Republic which is the part of the Ministry of Interior. (e.g. the Czech Red Cross receive 30 000 CZK, while Diaconia receive between 40 000 to 60 000 CZK per year).

3.3.2 Working group of ministry

Another form of cooperation between NGOs and Government Agencies in relation to Czech flood risk management is the Ministry of Interior working group for Volunteering and Emergency and Crisis situations. This group was established during the European Year of Volunteering. Attached to this working group are representatives from the Ministry of Interior, ADRA, the Czech Red Cross, Diaconia, Caritas and also from South Bohemian regional authority, ČEZ group, the General Directorate of Fire Rescue Service, Voluntary fire service and Masaryk University. One of the main goals of the group is encouraging mutual cooperation among NGOs, public administration and corporate volunteering.

3.3.3 Regional authorities

On a regional level, there are several form of cooperation with NGOs: Integration into the regional crisis plans, regional taskforce session, certification of public collections, regional PANEL and professionals' conferences

3.3.4 Integration into the regional crisis plans

NGOs are integrated into regional crisis plans in two ways: direct and indirect. Direct integration refers to an NGO role and participation in the regional crisis plan, while indirect integration occurs when an NGO are integrated through the Fire Rescue Service and the Fire Rescue Service is directly integrated into the regional crisis plan (see parts 'Emergency services of Integrated Rescue System')

An example of a direct integration is Diaconia. It is integrated into the crisis plan of Zlín Region. Examples of indirect integration are Caritas and ADRA, which are integrated into the crisis plans through Fire Rescue Service as an emergency service of Integrated Rescue System. NGOs also have opportunity to comment regional crisis plans (e.g. ADRA in South Moravia Region, Karlovy in the Vary Region, Liberec Region, and South Bohemia Region).

3.3.5 Regional taskforce session

Each region has its own task force for managing the crisis situations and in these task forces, NGOs representatives occasionally participate. We already have in the Czech Republic several examples of this type of participation, e.g. the Czech Red Cross, which takes part in task force sessions in many of regions and People in Need take part in Liberec Region task force.

Regardless, all NGOs have declared their willingness to take part in these sessions and all of them are convinced that they have something to offer. According to the NGO's, the sharing of information is a significant benefit of cooperation. Sharing information with NGO's could also be considered the biggest benefit on the municipal level (see part 'Municipal Authorities')

3.3.6 Certification of public collections

According to law on public collections [14], the Regional Governments of the Czech Republic or the Municipal Office of Prague give approval (certification) for the collection of money in the form of donations from the public. Money from public collections is an important part of financing participation of NGOs in Czech flood risk management.

3.3.7 Regional PANEL

On a regional level, is PANEL is another possible form of cooperation for NGOs, Unfortunately, PANELs are currently operating in only 3 regions from 14 (including Prague). Moreover even among these three regions there are significant differences. The South Moravia PANEL has signed contracts but there are doubts about practicality of the contracts working. In Ústí nad Labem region, PANEL is working but without the financial support of the regional government and in South Bohemia Region, PANEL enjoys significant financial support from regional government.

Each PANEL has its own unique structure. Among professionals from Czech emergency management there is relative consensus that the South Bohemia PANEL is currently the best available variant. In general one could say that PANEL is the working group for cooperation between NGOs and regional governments and for coordination of NGOs.

The South Bohemian Regional government has contracts with members of PANE. In 2011these included contracts with the Association of Firefighters of Bohemia, Moravia and Silesia, ADRA, Caritas and the Czech Red Cross) The regional government is providing financial assistance to these NGOs (470 000 CZK to Association of Firefighters of Bohemia, Moravia and Silesia, 720 000 CZK to Czech Red Cross, 80 000 CZK to ADRA and 80 000 CZK to Caritas, per annum in all cases.). If there is no emergency during the year, the NGOs use this money to finance their usual activities. NGOs also have the opportunity to request an increased financial assistance from the regional government if necessary. In the event of an emergency, an evaluation report is produced at the conclusion of the agency's emergency participation.

Through PANEL the NGOs have assigned roles to cope with the emergency. Czech Red Cross, for example, assumes the role of emergency coordinator; ADRA is providing seminars for volunteers, while the Association of Firefighters of Bohemia is cooperating with task force etc.

The South Bohemian PANEL is considered the best variant at present. Therefore the Czech Republic Ministry of the Interior are currently creating a model based on the South Bohemian PANEL structure for use as a standard in setting up other PANEL working group across other regions. That is not to say the model cannot be adapted to better suit local conditions but at present it is the optimal model. At present, it is expected that the Ministry of Interior Volunteering and Emergency and Crisis situations working group will coordinate these activities. One interesting opinion suggests that PANEL is the nonprofit sector's version of Integrated Rescue System.

3.3.8 Professionals' conferences

On the regional level, the conference platform is used as a means of introducing the problems, capabilities and needs of NGOs, local governments, Fire Rescue Service, volunteer firefighters etc. to each other. Some of these conferences in several regions are called 'Do we know about each other?' are initiated and supported by regional governments. South Bohemia region is an example of this.

3.4 Municipal authorities

Also on a municipal level, are several forms where cooperation with NGOs and their contribution were recognized – Integration into the municipal crisis plans, municipal taskforce session, municipal PANEL and helping the municipality to provide social service.

3.4.1 Integration into municipal crisis plans

As in case of regional governments, municipalities are cooperating with NGOs to develop or improve local crisis plans. An example of this direct form of integration is Hand for Help who have representation and a specified role in the crisis plan of municipality of Hrádek nad Nisou.

3.4.2 Municipal taskforce sessions

Several examples of participation by NGOs in Municipal task force sessions can be found. Czech Red Cross is taking part in task force sessions in many municipalities. People in Need participate in the municipalities of Nový Jíčín and Frýdlant. NGOs cooperate by sharing information with municipality governments and representation at the municipality government level - mayor, crisis managers etc.

This cooperation is quite important in the event that a flood occurs. Similar to the regional level, the participation of NGO's in flood task force session is having a positive effect and is to be recommended. For the same reason, it seems to be also more convenient for NGOs to delegate one representative to take part in task force session.

3.4.3 Municipal PANEL

Also at municipality level the PANEL cooperation is occurring (f. e. Děčín in Ústí nad Labem Region and Česká Lípa in Liberec Region), even this platform is still not common on the municipal level.

3.4.4 Helping the municipality with providing social services

An interesting example of NGOs – municipal government cooperation came as a result of the big flood in 1997. At that time, Czech Red Cross was building retirement houses in areas affected by flood. Subsequently most of them were given to municipal authorities.

3.5 Emergency service of integrated rescue system

This type of cooperation has several forms: cooperation at the moment of the emergency, contracts between the Fire Rescue Service and NGOs, cooperation at taskforce session, professionals' conferences and working group of ministry

3.5.1 Cooperation at the moment of emergency

Integrated Rescue System is based on emergency services of the Integrated Rescue System (Fire Rescue Service of the Czech Republic and Fire Units, Police of the Czech Republic and Medical Rescue Service). But according to law on the Integrated Rescue System [15], there are also other emergency services of the System and one of them are NGOs. All are considered elements of the system and all elements are expected to cooperate when dealing with an emergency.

It is obvious that in case of emergency the NGOs will have to cooperate with emergency services of Integrated Rescue System such as Fire Service and Fire Units, Police and Medical Rescue Service. In the event of a flood NGOs, particularly volunteers, must be able to communicate with the commander of Integrated Rescue System during the intervention.

3.5.2 Contracts between the fire rescue service and NGOs

The most important aspect of NGOs cooperation with emergency service of Integrated Rescue System in Czech flood risk management is contact with the Fire Rescue Services. Some NGOs (ADRA, Caritas) direct communication with the Fire Rescue Services and through them the NGOs are integrated into regional crisis plans.

Kavan [7] mentions three possible types of contracts between NGOs and the Fire Rescue Service in Integrated Rescue System of the Czech Republic: contract about the planned help on request; contract about the provision of personal or material assistance and contract about the cooperation.

3.5.3 Cooperation at taskforce session

Representatives from sectors of the Integrated Rescue System take part in Task Force meetings along with NGOs representatives where information sharing can occur.

3.5.4 Professionals' conferences and working group of ministry

Many representatives of Fire Rescue Services regularly take part in professional conferences about volunteering in emergency management both at the central level (provided by Ministry of the Interior) and the local level (provided by regional governments.) These conferences are by other levels of Government and relevant NGO representatives.

The Fire Rescue Service is represented in Working group of Ministry of Interior Volunteering and Emergency and Crisis situation.

3.6 Qua-volunteers

This category is a modification of Bullain and Toftisova's [10] category called Quango. According them, Quangos straddle the line between the state and organized civil society. In this newly established category the situation is a little different. In this case, qua-volunteers are voluntary firemen. They are usually members of volunteer firefighters NGOs, but at the same time they are also members of voluntary fire service.

They work on voluntary basis but as soon they decide to be a member of voluntary fire service units of the municipality, they have to obey orders. According to Czech legislature, a voluntary fire service unit is an organizational unit of the municipality and it is founded and financed by municipality. Their members also have some obligation toward municipality.

Voluntary fire service units are emergency service of the Integrated Rescue System. The voluntary firemen were placed in this category because of their specific characteristics. Apart from cooperation during a flood event while in their role as part of an Integrated Rescue System, the voluntary firemen are also share information with other NGO's during the professional conferences and participation in working groups such as Volunteering and Emergency and Crisis situation of Ministry of Interior.

3.7 Employment agency

In case of Czech flood risk management has appeared in the past one interesting example of this cooperation. Current Czech legislature requires that people registered at an Employment agency must fulfill certain terms and conditions. Without fulfilling the terms after some time, they are not receiving the unemployment support anymore. One of these terms and conditions is public service, which can be done also in some NGO. There is a one new thing according the law is now not possible to do this public service by volunteering according the law on volunteering. [16].

According the past legislative (before 2012) the situation was a little bit different. Public service was not obligatory for receiving social benefits, but it was an option for receiving extra money from the state. Also it was possible to do it by volunteering according the law on volunteering [16]. These people already took part in case of Caritas Czech Republic's participation in Czech flood risk management in 2010.

3.8 State or half-state companies

This category is not commonly considering when examining cooperation between NGOs with state. But some companies where the state holds a majority shareholding in the company of are contributing in this area. One such company is the energy conglomerate ČEZ. The state holds approximately 70% [17] of the shares in this company. ČEZ, not only donates for purposes of flood risk management, but is also supportive in other ways. For example, ČEZ has a quite sophisticated system of releasing employees for volunteering, including volunteering in flood risk management. It is also providing the graphics for promotional material on volunteering in the event of an emergency. A ČEZ representative is a member of working group of Ministry of Interior Volunteering and Emergency and Crisis situation therefore should be considered alongside NGOs in flood risk management.

Another example is Česká televize (ČT), a public television broadcaster in the Czech Republic. In February 2012, ČT introduced a project called crisis map of Czech. The project was launched in cooperation with the world volunteer network, The Standby Task Force. Prior to the launch of the project, discussions were held with NGOs specializing in humanitarian services for relevant data on which to base the project [18].

4 Conclusion

The study provides a unique point of view on the type of cooperation, in its various forms, between NGOs and Czech state. An overview of this type of cooperation could be helpful for other post-socialistic countries, faced with similar circumstances as the Czech Republic. That is trying to build civil society and infrastructure from the void left by the communist regime and the growing problem of floods.

It could provide another point of view for consideration by foreign NGOs and their states on how the system works in this area in the Czech Republic for countries with similar pasts. Further research could investigate what factors influence success or otherwise of the cooperative process between NGO's and Czech Government Agencies as well as identify what successful cooperation means. Why this is important is obvious.

If other post-socialistic countries want to gain understanding of the Czech experience in this area for application in their own country, it is not enough just to say that we moved from point A to point B but also how. It is not easy to answer the question of how but it should be possible to identify factors of successful cooperation, or factors which impede cooperation. These factors should be analyzed and compared and universal truths draw out if they are there.

Moreover, this overview of all forms of NGOs – State cooperation related to Czech flood risk management could be helpful for Czech state and NGOs because cooperation is so variable at all levels across the country. From past Czech experiences one can assume that floods will occur whether or not the region affected has a well-developed system or not.

The study also suggests areas that require further research particularly the economic justification of financial support of the PANEL platform, projects from European funds as additional financial sources, status of voluntary firemen in the Czech Republic, well as the need for deeper research into the participation of state companies in some kind of emergency management.

References

[1] Palubinskas, G. T. (2003) 'Democratization: The Development of Nongovernmental Organizations (NGOs) in Central and Eastern Europe', Public Administration and Management, vol. 8 No. 3.

[2] Frič, P., Deverová, L., Pajac, P., Šilhánová, H. (1998) 'Defining the Nonprofit Sector: The Czech Republic.' Working Papers of the Johns Hopkins Comparative Nonprofit Sector Project, No. 27, Baltimore: The Johns Hopkins Institute for Policy Studies.

[3] Brhlíková, P (2004) 'The Nonprofit Sector in the Czech Republic.', Discussion Paper No. 2004-128, Praha: CERGE-EI, Charles University.

[4] Hyanek, V., Pospišil, M. (2007) Country-Specific Situation of the Nonprofit Sector in the Czech Republic, Brno.

[5] Rektořík, J. a kol. (2004) Krizový management ve veřejné správě. Praha: Ekopress. s. 250. ISBN 80-86119-83-1

[6] Antušák, E., Kopecký, A. (2005) Krizový management. Úvod do teorie.

[7] Kavan Štěpán. Ochrana obyvatelstva I. Vysoká škola evropských a regionálních studií, České Budějovice, 2011, 109 p.

[8] Kumar, P. (2005) NGO Coordination in Humanitarian Action: The Case of the Czech Floods of August 2002, Massachusetts Institute of Technology.

[9] Rektořík, J., Šelešovský J., Bakoš, E., Dostál. J., Furová, L., Šrámková, T., Smutný, M. (2011) 'Zpráva z výzkumného projektu "Ekonomické a finanční dopady živelních pohrom"', Brno.

[10] Bullain, N., Toftisova R. (2005) 'A Comparative Analysis of European Policies and Practices of NGO-Government Cooperation', The International Journal of Not-for-Profit Law, vol 7, No. 4.
[11] Government Council is the Council for Non-Governmental /Non-Profit Organisations. Basic information. http://www.vlada.cz/cz/ppov/rnno/zakladni-informace-767/
[12] Government Council is the Council for Non-Governmental /Non-Profit Organisations. Members of the council http://www.vlada.cz/cz/ppov/rnno/clenove/default.htm
[13] Ministry of the Interior of the Czech Republic. Voluntary service. http://www.mvcr.cz/clanek/dobrovolnicka-sluzba-500539.aspx?q=Y2hudW09Mw%3D%3D
[14] Law no. 117/2001 on public collections
[15] Law No. 239/2000 on the Integrated Rescue System § 4 (2)
[16] Ministry of the Interior of the Czech Republic. Vyjádření ke změnám v souvislosti se Sociální reformou I. http://www.mvcr.cz/soubor/dobrovolnicka-sluzba-x-verejna-sluzba-2012-doc.aspx
[17] ČEZ annual report (2012).
[18] Crisis map of Czech http://www.ceskatelevize.cz/ct24/media-a-technologie/164631-krizova-mapa-ceska-divaci-v-roli-novinaru-pri-krizovych-situacich/

The utilization of an abandoned mining pond as a retention pond in dealing with stormwater runoff

H. Takaijudin, A. H. A. Shahir & A. M. Hashim
Civil Engineering Department, Universiti Teknologi Petronas, Perak, Malaysia

Abstract

Most abandoned ponds can be upgraded as detention or retention ponds. However, some of the abandoned mining ponds have been used in an improper way such as for illegal landfill sites. The selected ponds in this study fulfil the characteristics as retention ponds in terms of location, pond capacity, inlet and outlet discharges and the facilities installation. Design rainfall was obtained from MASMA manual using Bagan Serai design storm coefficient. The peak flow rate for the drainage system was designed in 10 years ARI due to having the development of housing and industrial areas. The results show that the peak runoff by using design rainfall is much higher than the peak flow from current rainfall. The volume of the pond has been designed for 10, 50 and 100 years ARI (Average Recurrence Interval). The storage can be exceeded up to 130000m^3 whereas the availability of this natural pond is up to 156000m^3. Thus it is concluded that the capacity of the ex-mining pond is sufficient to cater for the runoff from catchment areas. The abandoned mining pond has the potential to be developed as part of the infrastructure in future.
Keywords: ex-mining pond, retention pond, peak discharges.

1 Introduction

Stormwater management tools such as Low Impact Development (LID) or Water Sensitive Urban Design (WSUD) offer a sustainable solution to decentralized stormwater management such as the wet pond which can be implemented in watershed scale [1]. A series of ponds can promote flow attenuation with the installation of outlet structures such as V-Notch weir and orifices [2]. Retention ponds offer water quality and flood control by the natural process in the ponds.

WIT Transactions on Ecology and The Environment, Vol 159, © 2012 WIT Press
www.witpress.com, ISSN 1743-3541 (on-line)
doi:10.2495/FRIAR120031

They provide the most effective storm water treatment practices at removing pollutants. The water contained in the permanent pool mixes with and dilutes the initial polluted runoff from heavy rainfall events. The retention pond also helps plants, algae, and bacteria consume the pollutant for their food and energy to help their growth. Thus, it will reduce the total mass of pollutants released downstream. The abandoned mining pond can be considered for the community event as regional pond facilities. It has the potential to be upgraded as a series of wet pond in some developed areas due to site constraint and economic considerations [2]. The current practice in Malaysia is that the developers and engineers prefer retention facilities to handle runoff from development activities. They can upgrade abandoned ponds to cater for the runoff from residential areas. Most of the natural processes can be adopted using this method. It functions as storage and releases the flow gradually through outlet control structures [3].

Since 1820, tin mining was the most active industry in Malaysia. The mining started after the arrival of hardworking Chinese immigrants, mostly Cantonese and Hakka, who provided the manpower. Tin mining mainly occurred in the West Peninsular of Malaysia and as a result of this activity in that area ex-mining land has covered approximately 113700ha since the 1930s [4]. The alluvial deposited soil mainly the soil type where tin can be found. The sand is infertile and the area has a harsh microclimate, whereas the slime is more fertile but normally water logged [5]. The state of Perak was the eldest mining pond site. Taiping is the first town where tin was being commercialized and by 1883, Malaysia had become the largest tin producer in the world. It supplied around 55% of the world's tin by the end of the 19th Century, compared with 30% in 1992.

During the 1990s, many mining ponds stopped their operations and were abandoned for a long time. Only 4730ha (4.2%) of the ex-mining land had been utilized for agriculture and 5.5% for other uses [4]. Although the government included the Act of ex-mining ponds which stated that the rehabilitation of ex-mining ponds would be used for certain plans. Unfortunately, the plans are still not clear. Narrowing it down to Perak state itself, the mining pond, an abandoned pond is mainly a waste area. Some of the ponds have been deposited with solid waste and become illegal landfill sites. In terms of cost-benefit, it is predicted to be the best detention facility due to natural causes.

The quantity and quality of water needs to be considered for design and analysis of an ex-mining pond. The 10-year (ARI) design storm event is recommended for minor conditions while 50-year ARI for major conditions in controlling peak flow and water quality [6]. For this reason, this paper presents the investigation of the availability of the pond storage to cater for surface water which comes from mixed developments, particularly from residential and industrial areas.

2 Project area description

The study area covers Lot 5057, Lot 3873, Lot 7805 and Lot 11979 of Kamunting Raya, Taiping, Perak. It is located in the Larut Matang and Selama territory and situated about 10km from Taiping Utara PLUS Toll. The proposed site is still on a development program. The developed area consists of industrial areas which are mainly factories and training centres and some residential areas. The area is about 200 hectares and almost flat ground level. The proposed site has three abandoned mining ponds. Fig. 1 illustrates the satellite image of ex-mining ponds which are located in the middle of the developed area.

The study has selected one pond which receives the runoff from residential areas and some industrial areas. Fig. 2 demonstrates the drainage network model

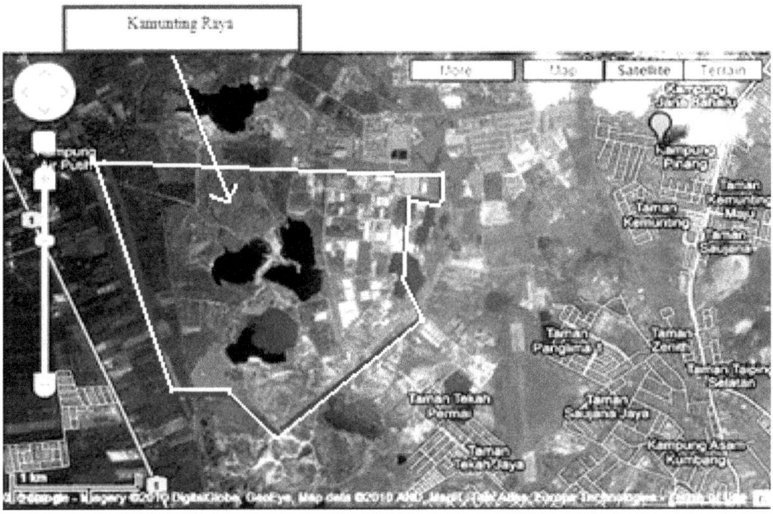

Figure 1: Satellite view of Kamunting Raya area.

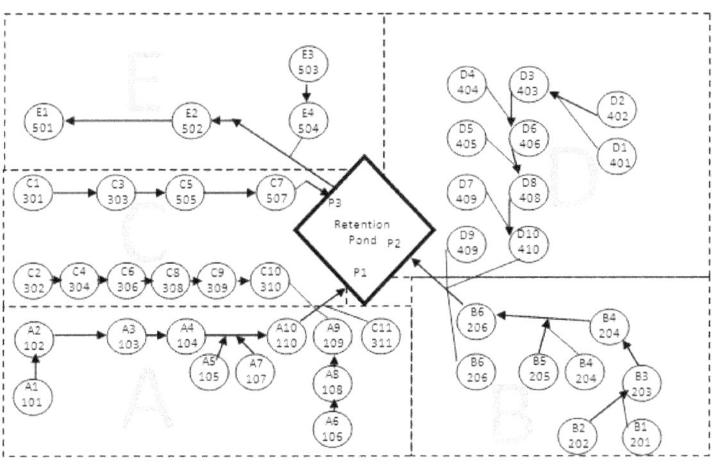

Figure 2: Drainage network with the integrated pond.

using MIDUSS software at the catchment areas. The drainage areas of Kamunting Raya are connected to each other. There are 3 inlets that cater for runoff from catchment A (59.38 ha), B (29.40 ha), C (58.42 ha) and D (56.11 ha). There is one outlet from the retention pond that is discharged to the block drain e2-e5 in catchment E. However, runoff catchment E has been discharged to another pond which is located at the downstream part. The main drain will cater for the runoff stormwater for the larger area. P1, P2 and P3 are the inlets to the retention pond. The final discharge for the overall area is at Sungai Air Putih, Kamunting, Perak.

3 Methodology

3.1 Data collection

Data has been collected from various local authorities, developers and consultants for this particular site. 20-years rainfall data from 1989 to 2009 were recorded at JKR Taiping rain gauge. The catchment areas of the site were provided by DID. For specific land in Malaysia, local authority is given the power to control and manage the land. Therefore, for this project some local authorities were consulted for the legal data. Other than that, for the real experience of the site conditions, the consultants and developers have to be consulted.

3.2 Design criterion

Rainfall is the main contributor of natural water resources. It flows as surface runoff, infiltrates into the soil and will evaporate. In the design, the rainfall intensity is taken into consideration as solely as the surface runoff. Drainage network was designed for 10 years ARI while the pond system for 100 years ARI. The contribution in design stages is in terms of the rainfall intensity. Intensity (*I*) can be calculated using eqn (1):

$$ln({}^{R}I_{t}) = a + bln(t) + c(ln(t))^{2} + d(ln(t))^{3} \qquad (1)$$

where,

${}^{R}I_{t}$ = the average rainfall intensity (mm/hr) for ARI and duration

R = average return interval (years)

t = duration (minutes)

a to *d* are fitting constants dependent on ARI for Bagan Serai area.

The next step will be on the determination of peak flow by applying the Rational Method. This method is used to estimate the peak flow for each conveyance system. This method was applied to the manual calculations (spreadsheet) and MIDUSS software. Using the same method, the result was expected to give significant output with lower percentage difference. Specific geometry of drainage was selected in order to cater for the peak flow of the drainage system. Once the drainage design was completed, the pond was designed using Swinburne Method as recommended by MASMA [2]. In this method, the permissible site discharge (PSD) and Site Storage Requirement were determined. However, MIDUSS configuration has provided a different approach in which based on the final discharge from the inlet pond, the pond was designed with certain geometry. It will provide storage routing, the inflow and the outflow hydrograph of the pond.

4 Results and discussion

4.1 Physical evaluation of ex-mining pond

A mining pond is a natural pond. The depth is greater than other man-made ponds. The depth of the pond is about 15m. Therefore, a lot of consideration is carried out. General limitations adapted from MASMA Volume 8 are analyzed to evaluate the physical characteristic of the pond. Table 1 lists the requirements.

Table 1: The physical criteria evaluation of retention pond (adapted from MASMA 2000).

Criteria	MASMA Requirement	Kamunting Raya Ex-Mining Pond
1) Soil Suitability	• No limitation on soil infiltration rate but a minimum rate of 13 mm/hr is recommended. • Soil with greater clay content or 40% greater silt/clay content shall not be used.	• No limitation of soil infiltration. • The soil is consisting of sandy soil. • The system use is natural and no fill material is utilized.
2) Depth to Bedrock, Water Table,	• All facilities shall be located at least 1.5m above the seasonal high ground water mark, bedrock (or hardpan) and/or impermeable layer.	• These criteria have been implemented. The level of the building is increase by 2m.
3) Proximity to Drinking Water Wells, Septic Tanks, Drain fields, Building Foundations, Structures and Property Lines	• Infiltration facilities on commercial and industrial sites at least 35m from drinking water wells, septic tanks or drain fields and spring used for public drinking water supplies. • Infiltration facilities should be situated at least 7m down slope and 50m from building foundations.	• Building is located more than 100 m from the retention pond. • This system is compulsory according to building bye-law. • The planner has been planned the building placement before construction.
4) Land Slope	• Avoid application of infiltration practices on a steep grade. • Infiltration facilities can be located on slopes up to 15% only.	• The site area is almost flat whereby the gradient is approximately 0.001.
5) Drainage Area	• Infiltration BMPs are limited in their ability to accept flows from larger drainage areas. • Infiltration basin – maximum of 15 hectares	• The overall catchment area is about 200 hectare. • The catchment area has been divided according to the inlet.

from MASMA and the description from the abandoned mining pond observation in order to comply with the requirements.

4.2 Hydrologic analysis

4.2.1 Rainfall analysis

For this project, design rainfall is used as suggested by MASMA. The storm duration use is 30 and 60 minutes as suggested. The adjustment factor is used for the time of concentration below 30 minutes. Bagan Serai is chosen for the Design Rainfall as it is the nearest place from Taiping. The drainage network has been designed for 10 years ARI because of the mixed development. The pond has been designed for 50 and 100 years ARI return period. The Malaysian version of MIDUSS has an extra item in the Hydrology menu for design storms for use in Malaysia. Because of the input which uses the same design rainfall, the return period can be chosen to be the same as manual calculation using 10, 50 and 100 years ARI. The default time of concentration is 30 minutes. MIDUSS application is applicable for 2, 5, 10, 20, 50 and 100 years ARI. The design rainfall is used rather than current time series rainfall for MIDUSS application.

The advantage of the time series rainfall data analysis is that the depth is clearly shown according to the return period. However, the analysis method can be too straight forward and the result of the Intensity is much lower than the design rainfall. In addition, 24 hour rainfall duration is too long. Therefore, more comprehensive statistical analysis has to be carried out based on the data. The method used for this project has limitations and further calculations have to be made.

4.2.2 Peak runoff and pond capacity

In using manual calculation, the drainage design is adapted using MASMA manual. Two (2) types of Rainfall intensity are used for these design purposes. The first design is using the design storm which is recommended by MASMA and secondly the current rainfall data which has been calculated previously. For a mixed development, the return period of 10 years is required. The time of concentration which has been investigated earlier needed in the calculation using design storm. In the current rainfall calculation, the rainfall intensity is being used directly in the calculation.

Next, the rational method is used for calculating the peak discharge of the drain. The peak discharge is compared to the given drain capacity to determine the real size of the drain. The connected flow of the catchment is taken into account of which in Catchment A there is a link of catchment. The pattern is that the first catchment will have a smaller drain size compared to the last catchment which links to the retention pond. MIDUSS provide different approaches where the hydrograph technique is applied in the model. A percentage of pervious and impervious catchment is specified. The drainage system is design for post development. The percentage of impervious was taken as up to 50% for Kamunting Raya area. Fig. 3 shows the runoff hydrograph for pervious and impervious areas.

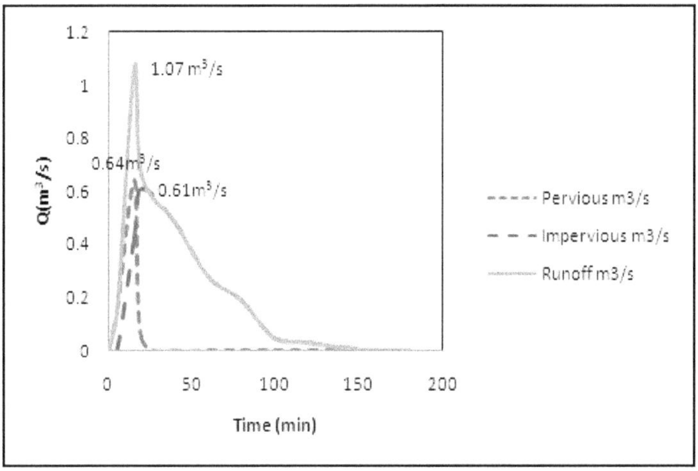

Figure 3: Runoff hydrograph for impervious and pervious areas.

The peak discharges for each drain and the outlet structures have been compared using MASMA manual calculation and MIDUSS software. Figs. 4 and 5 illustrate the difference between values of peak runoff for the selected drain using these two methods. The percentage difference is approximately 4-15% error by comparing the result of manual calculation and MIDUSS.

For the pond design the volume is calculated based on Permissible Storage Discharge (PSD), the Site Storage Requirement (SSR) and the allowable storage of the existing pond. From the results obtained, the abandoned mining pond is sufficient to cater for the stormwater for the overall catchment. Table 2 lists the capacity of ex-mining pond system for this catchment area. The inlet of the discharge is assumed to be in one inlet. The overall catchment gives a

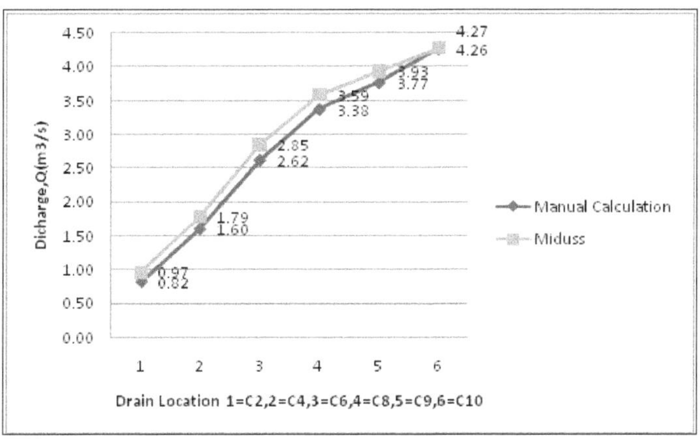

Figure 4: The comparison of peak flow for drainage link C2-C10.

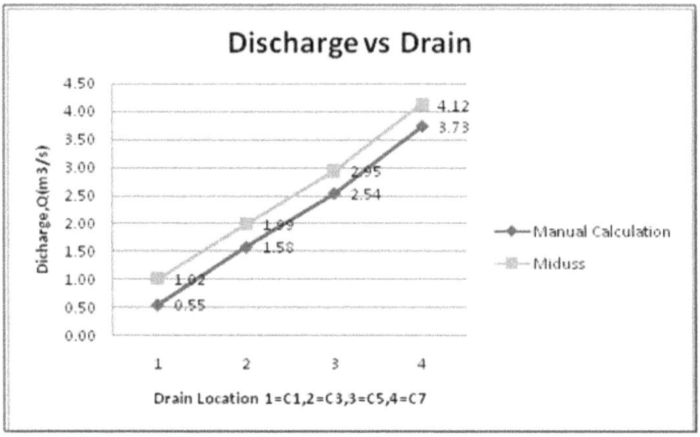

Figure 5: The comparison of peak flow for drainage link C1-C7.

Table 2: Abandoned mining pond capacity.

	Abandoned Mining Pond
PSD_{10} (l/s)	48346.81
PSD_{100} (l/s)	67964.29
SSR_{10} (m^3)	90435
SSR_{100} (m^3)	126637
Allowable Storage(m^3)	156240
Average Depth of Pond (m)	5.0
Average Platform Level (m)	GL 8.5464
Average Water Level (m)	IL 7.7297

conventional number of peak discharges for the pond. The pond is designed solely based on the peak discharge that has been calculated. Therefore, the pond volume is higher but still under volume when compare with the volume of the abandoned mining pond. MIDUSS also provide pond design (Fig. 6) and outlet structure such as weir and orifice. There are 2 inflows that produced a higher amount of discharge for 50 year ARI and 100 year ARI as in Figs. 7 and 8 where the allowable outflows are 25.2% to 65% respectively, lesser than the inflow. Fig. 9 illustrates the dimension of one of the natural ponds in the study site. However, the depth might vary due to it being a natural pond. The estimated depth of this pond is about 10-15m depth approximately. The final discharge is conveyed to the main stream of Sungai Air Putih.

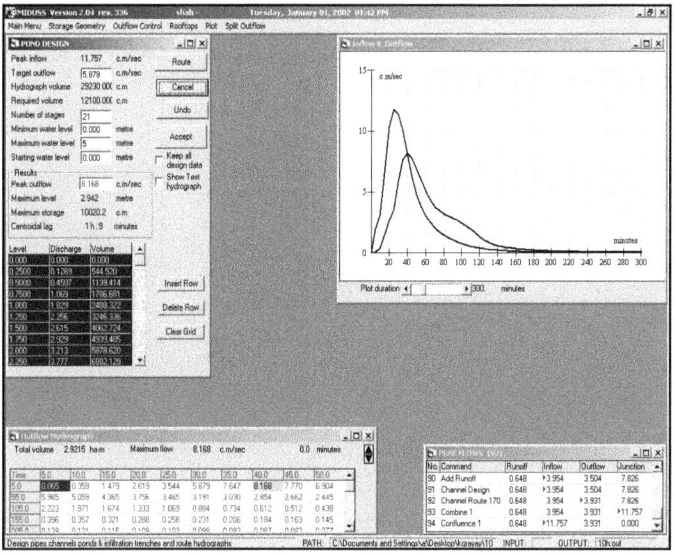

Figure 6: Pond design features by using MIDUSS application.

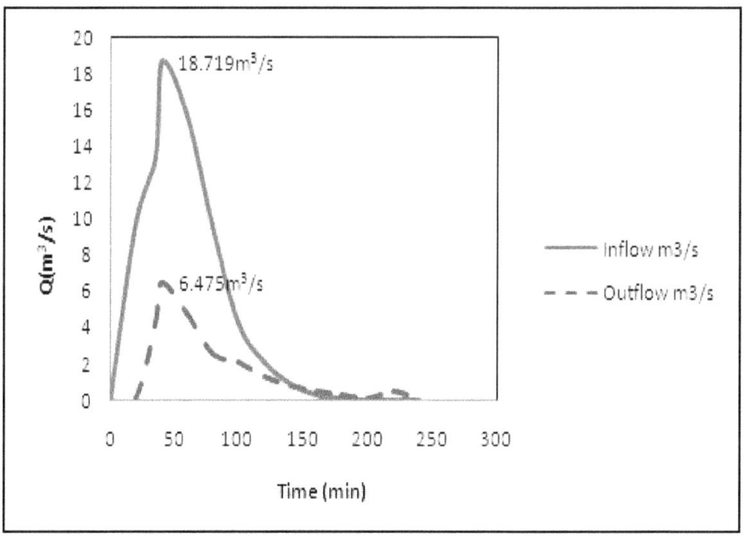

Figure 7: Inflow and outflow for pond system in 50 years ARI.

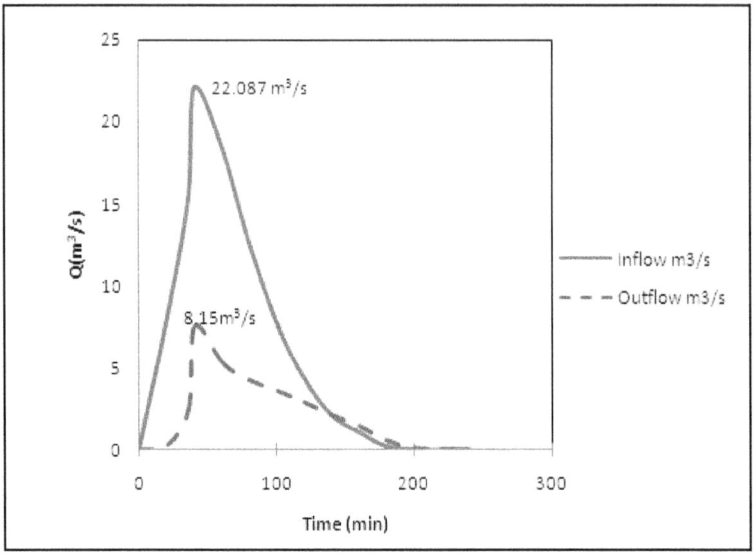

Figure 8: Inflow and outflow for pond system in 100 years ARI.

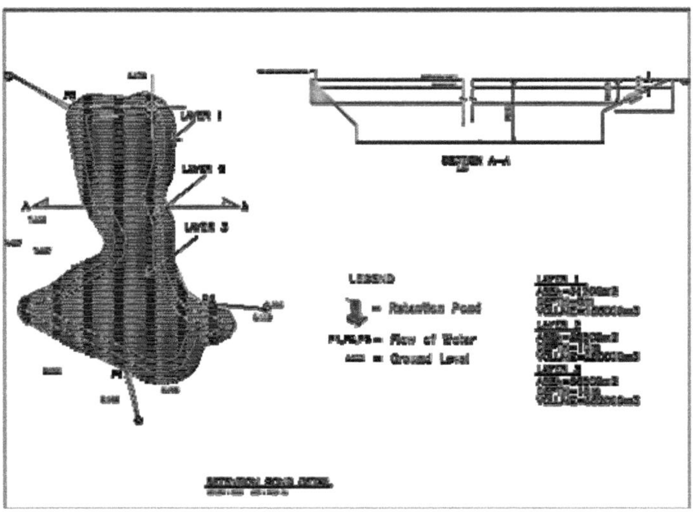

Figure 9: Ex-mining pond layout using ACAD.

5 Conclusion

The study confirms that the abandoned pond is suitable to be upgraded as a retention pond due to the location of the pond as it is downstream of the catchment area. Thus all peak runoff can be channelled easily using gravity flow.

In addition, the capacity of the existing pond is sufficient to cater for the design storm discharge.

The result from MIDUSS and manual calculation produced a very good simulation and it is applicable for analysis of the series of ponds.

The capacity of this pond is that it is sufficient to cater for the higher volume up to a design life of 100 year ARI.

Abandoned mining ponds are usually viewed negatively. It is timely to shift this negative perception to a positive view where with proper planning, proper design of storm water management system, and suitable layout for land use.

Ex-mining ponds will soon be prime real estate masterpieces for the future while generating revenue for the state government.

....With the information available in this case study it is recommended that further study should be made in term of the water quality aspect especially for stormwater treatment.

References

[1] Roy, A.H., Wenger, S.J., Fletcher, T.D., Walsh, C.J., Ladson, A.R., Shuster, W.D., Thurston, H.W., Brown, R., Impediments and Solutions to Sustainable, Water-Scale Urban Stormwater Management: Lessons from Australia and the United States. *Environmental Management*, **42(2)**, pp. 344-359,2008.

[2] Department of Drainage and Irrigation (DID), On-site and Community Retention (Chapter 21) and Regional Retention (Chapter 22), *Urban Stormwater Management Manual for Malaysia,* **8**, 2000.

[3] Takaijudin, H., Hashim, A.M., Vallyutham, K., Shahir, A.H.A., Ghazali, A.H., Implementation of Urban Stormwater Management Practices in Malaysia – A Review, *1st International Conference of Sustainable Building and Infrastructure 2010*, Kuala Lumpur, 2010.

[4] Chan, Y.K., The Mining Land – An Overview of the Current Situation in Peninsular Malaysia. *Paper presented at the Seminar on Ex-Mining Land and BRIS Soil: prospects and Profits*, Kuala Lumpur, p. 17, 1990.

Ho, W.M., Ang, L.H., and Lee D.K., Assessment of Pb Uptake, Translocation and Immobilization in Kenaf (Hibiscus cannabinus L.) for phytoremediation of sand tailings, *Journal of Environmental Sciences*, **20**, pp. 1341-1347, 2008.

[5] Chang, C.K., Ghani, A.A, Zakaria, N.A., Yusof, M.F., Ayub, K.R., Lai, S.H., Hasan, A.M. M., Ainan, A., Rehabilitation of Ex-Mining Pond and Existing Wetland for Integrated Stormwater Management, *11th International Conference on Urban Drainage*, Edinburgh, Scotland, UK, pp. 1-10, 2008.

Section 2
Flood risk vulnerability

Flood-risk assessment of the dense downtown in Fukuoka City, Japan

H. Hashimoto[1] & Y. Nonaka[2]
[1]*Department of Civil Engineering, Kyushu University, Japan*
[2]*Osaka hdqrs., CTI Engineering Co. Ltd., Osaka, Japan*

Abstract

The flood risk of the Naka River is much higher than the other rivers in the Fukuoka Metropolitan area. The Naka River watershed has the dense downtown along the lower river reach. If the Naka River and its tributaries would overflow in the lower river reach, significant economic loss would be caused. The urban flood simulation detects areas of higher flood-risk in the downtown. The urban flood simulation is divided into two parts; one is one-dimensional flood flow simulation along the lower Naka River reach and the other is two-dimensional flood flow simulation on the downtown area. Furthermore, the latter is subdivided into two sub-parts; one is two-dimensional flood flow simulation for the ground level and the other is that for the underground space. The land use in the downtown is complicated. The land use in the downtown is incorporated into the simulation.
Keywords: urban flood, flood risk, urban flood simulation, underground space.

1 Introduction

On June 29, 1999 and July 19, 2003, short-time heavy rains occurred in the Mikasa River watershed of the Fukuoka metropolitan area and caused the river to overflow at some sections. Since these events, Fukuoka Prefecture Government has renovated the Mikasa River. As a result, the flood risk of the Mikasa River becomes very low. On the other hand, that of the other rivers such as the Naka River becomes much higher. If the rainfall of same magnitude as that on June 29, 1999 or July 19, 2003 would occur in the Naka River watershed, the Naka River and its tributaries would overflow. The Naka River watershed has the densest downtown along the lower river reach (fig. 1). In this downtown

WIT Transactions on Ecology and The Environment, Vol 159, © 2012 WIT Press
www.witpress.com, ISSN 1743-3541 (on-line)
doi:10.2495/FRIAR120041

there are many office buildings, some department stores, a few hospitals, City Government office buildings and the underground spaces utilized for the shopping mall, subway stations and parking garages. If the Naka River and its tributaries would overflow along the lower river reach, significant economic loss would be caused. Hashimoto *et al.* [1] made research on the flood-risk assessment of the lower Naka River reach. They presented the one-dimensional flood simulation for the lower Naka River reach and assessed the flood risk of the downtown along the lower river reach. However, the flood risk of the underground spaces was not assessed.

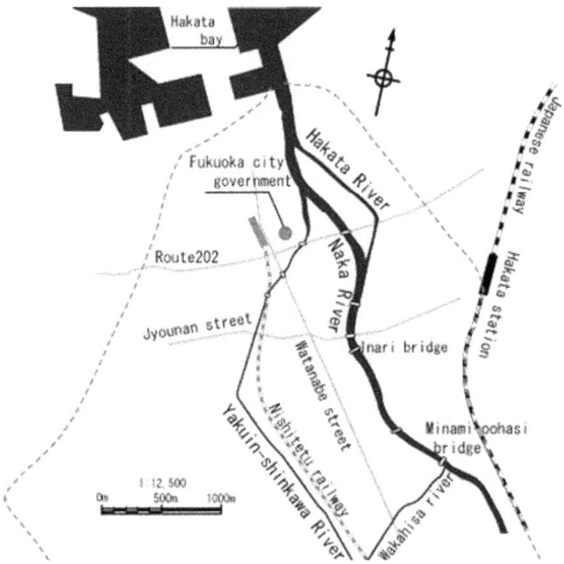

Figure 1: Plan view of the lower Naka River reach and its tributaries.

The purpose of the present study is to simulate flood flow and detect inundation areas in the dense downtown of the Naka River watershed under the same rainfall condition as that on June 29, 1999. The simulation of flood flow is divided into two parts; one is one-dimensional flood flow simulation in the lower Naka River reach and the other is two-dimensional flood flow simulation on the downtown area. Furthermore, the latter is subdivided into two sub-parts; one is two-dimensional flood flow simulation on the ground level and the other is that on the subway station concourse (the 1st basement). The present study is an extension of the previous work [1].

There are several urban flood simulation models (e.g. Fukuoka *et al.* [2], Toda *et al.* [3]; Haider *et al.* [4]; Hashimoto *et al.* [5]; Hashimoto and Park [6]). The urban flood simulation model of Hashimoto *et al.* [5] and Hashimoto and Park [6] is used in this study. This model considers area density of buildings and houses as the effect of their high density on flood flow. The land use in the downtown is incorporated into the simulation by using the concept of the area density.

2 Studied rivers and areas

The Naka River is 35km long and its watershed is 124km^2 wide. The upstream area is mountainous region with two dams, and the downstream area is a centre of business, shopping and amusement in the Fukuoka Metropolitan area. Urbanization is significant in the whole watershed except the mountainous areas. Therefore, flood control and measure is important in this river.

Studied reach of the Naka River is 3.7km long from Minami-Ohashi bridge to the river mouth (fig. 1). This reach has two tributaries. The Yakuin-shinkawa River enters the main river from the left-hand side at the place 0.88km distant from the river mouth. The Hakata River separates from the main river to the right-hand side at the place 1.72km distant from the river mouth. Moreover it enters the main river at the place 0.5km distant from the river mouth. The left-hand side area of the Yakuin-shinkawa River, called 'Tenjin district', has Fukuoka City Government office, an underground shopping mall, subway stations, department stores, hospitals etc.

3 Scenario of rainfall and sea level change

Rainfall scenario adopted in the present simulation is the measurement of rainfall at the Fukuoka Meterological Observatory Station on June 29, 1999. The same rainfall condition as the measurement is assumed in the whole Naka River watershed. Fig. 2 shows hourly rainfall situation on June 29, 1999 and runoff analysis at Minami-Ohashi bridge from the upstream Naka River watershed. Intermittent rainfall continued from June 28. Afterwards, continuous and heavy rainfall hit the Fukuoka Metropolitan area from 6 to 11 o'clock on June 29. The hourly rainfall rapidly increased at 7 o'clock, and then reached the peak value of 77 mm/hour from 8 to 9 o'clock. Cumulative rainfall was 164mm from 6 to 11 o'clock. This rainfall was typical short-time heavy rainfall.

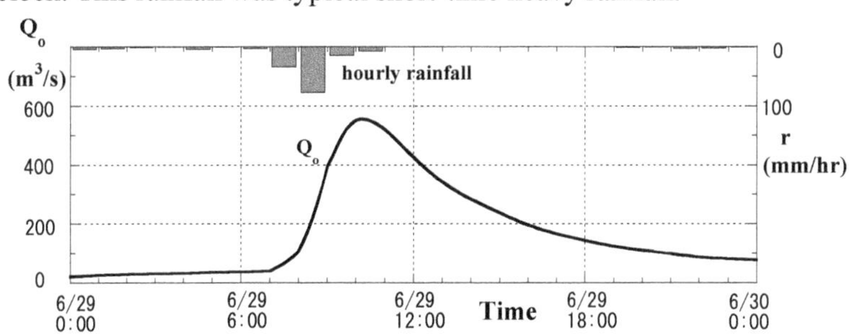

Figure 2: Rainfall scenario and flood discharge.

The boundary condition at the downstream end of the studied river reach is sea level change at the river mouth i.e. Hakata Bay. On the other hand, the boundary condition at the upstream end of the studied river reach is flood discharge Q_o at Minami-Ohashi bridge. The flood discharge hydrograph at Minami-Ohashi bridge has been already obtained with runoff analysis of the upper watershed by CTI Engineering Co. Ltd (fig. 2).

4 Review on the one-dimensional flood simulation model for the lower Naka River reach proposed by Hashimoto *et al.* [1]

Hashimoto *et al.* [1] proposed the one-dimensional flood simulation model for the lower Naka River reach. In this section we describe the basic equations and the boundary conditions for the simulation. The studied river reach is the lower Naka River reach and its tributaries, i.e. the Yakuin-shinkawa and the Hakata River.

4.1 Basic equations

One-dimensional flood flow analysis requires momentum and continuity equations as basic equations, and boundary and initial conditions. We consider lateral overflow of floodwater and lateral inflow of rain water through riversides. Lateral overflow discharge can be estimated by the overflow equation. Lateral inflow of rain water from each sub-basin area can be given by the rational equation. Flood flow discharge Q and depth h are unknown variables in the basic equations. Flood flow in the rivers can be treated as one-dimensional gradually varied flow. The cross sections of the rivers are almost rectangular. The basic equations are as follows:

Momentum equation:

$$\frac{\partial Q}{\partial t} + \frac{\partial (vQ)}{\partial x} = -gBh\frac{\partial (h+z)}{\partial x} - \frac{(B+2h)}{\varphi^2} \cdot |v|v - \beta q_{out} v \tag{1}$$

Continuity equation:

$$B\frac{\partial (h+z)}{\partial t} + \frac{\partial Q}{\partial x} = -q_{out} + q_{in} \tag{2}$$

Resistance rule:

$$\varphi = \frac{v}{u_*} = R^{1/6} \big/ (n\sqrt{g}) \tag{3}$$

Overflow and inflow discharge through riversides:

$$q_{out} = 0.35 h_{out} \sqrt{2gh_{out}} (1-\lambda_{out}) \xi_c \tag{4}$$

$$q_{in} = Q_i / \Delta L_i = \frac{f}{3.6} r(t) A_i / \Delta L_i \tag{5}$$

where t: time, x: distance measured in the downstream direction from the upstream end of the studied river reach, Q: discharge, h: flow depth, v: depth-averaged velocity, R: hydraulic radius defined as $R = Bh/(B+2h)$, z: river bed elevation measured vertically upward to the x-axis, B: river width, φ: nondimensional average velocity, n: Manning's roughness coefficient, q_{out}: overflow (i.e. outflow) discharge per unit riverside length, h_{out}: overflow depth on riverside bank, λ_{out}: area density of buildings on riverside bank, β=1.5: coefficient related to momentum transfer due to overflow (Hashimoto *et al.* [5]), ξ_c: correction factor related to energy loss, A_i: sub-basin area (km^2) from which rain water flows into river, Q_i: lateral inflow discharge from sub-basin of A_i, $r(t)$: rainfall intensity (mm/hour), ΔL_i: distance between neighbouring river sections into which rain water flows from each sub-basin area and f: runoff coefficient of rain water from each sub-basin area.

Overflow discharge eqn. (4) considers dense houses and buildings along riversides by using the concept of their area density.

4.2 Flood flow in the Naka River

We consider lateral overflow (i.e. outflow) q_{out} of floodwater from the river but neglect lateral inflow q_{in} of rain water from each sub-basin area because of its minor value compared with river flow discharge Q.

The boundary conditions of the computation are flood flow discharge $Q_o(t)$ at the upstream end (Minami-Ohashi bridge) and water level $H_m(t)$ at the downstream end (the river mouth). Flood flow discharge has been already evaluated by the runoff analysis under the initial wet land condition. Since the case of initial dry land condition yields smaller runoff discharge, we discuss the case of the initial wet land condition as the more dangerous one. Water level at the river mouth can be approximated by the measurement of sea level change at Hakata Bay.

$$Q = Q_o(t) \quad \text{at} \quad x = 0 \text{ km (Minami-Ohashi bridge)} \tag{6}$$

$$H = H_m(t) \quad \text{at} \quad x = 3.7 \text{ km (River mouth)} \tag{7}$$

Initial time of the computation is 0:00 on June 29, 1999. Therefore, the initial conditions are the values of flood flow discharge $Q_o(t)$ and sea level $H_m(t)$ at 0:00 on June 29, 1999. These values are obtained from eqns. (6) and (7). Using these boundary conditions of flood flow discharge and sea level, we can obtain initial flow depth at any positions along the rivers with non-uniform flow analysis.

4.3 Confluence of the Yakuin-Shinkawa and the Naka River

Yakuin-shinkawa River flows into the Naka River at the confluence place ($x = x_c$ = 2.82km). The inflow of the tributary to the main river can be regarded as lateral inflow for the main river. Indicating flow discharge of the Yakuin-shinkawa River by Q_y, we can express the flow of the Yakuin-shinkawa River as the lateral inflow discharge Q_y from the left-hand side of $x_c \sim x_c + \Delta x$. Q_y becomes negative when reverse flow occurs from the Naka to the Yakuin-Shinkawa River. Therefore, the continuity equation for the river reach $x = x_c \sim x_c + \Delta x$ can be rewritten as

$$B\frac{\partial(h+z)}{\partial t}+\frac{\partial Q}{\partial x}=-q_{out}+Q_y/\Delta x \text{ for } x=x_c \sim x_c+\Delta x \tag{8}$$

Here we consider q_{out} through its right-hand side but neglect q_{out} through its left-hand side, because the Yakuin-shinkawa River flows into the Naka River from its left-hand side.

4.4 Separation of the Naka and the Hakata River and their confluence

The Hakata River separates from the Naka River at $x=x_d$ and returns to the Naka River at $x=x_{ch}$.. Denoting flow discharge of the Hakata River by Q_h, we can obtain the continuity equations in the same way as the above discussion.

4.5 Flood flow in the Yakuin-shinkawa River

We consider lateral overflow of floodwater and lateral inflow of rain water through riversides. Here, simultaneous overflow and inflow through same riverside never occurs. The watershed of the river can be divided into several sub-basins.

As the boundary conditions, flow discharge can be given at the upstream end of the Yakuin-shinkawa River and water level at the confluence of the Yakuin-shinkawa and the Naka River. Therefore, we have $Q_y = 0$ at the upstream end of the Yakuin-shinkawa River, and $H_y = H_n$ at its downstream end, where H_y: water level of the Yakuin-shinkawa River and H_n: water level of the Naka River.

There is usually no flow in the upper river reach and initially less rainfall. Therefore, initial flow discharge is Q_y=0. On the other hand, initial flow depth of the Yakuin-shinkawa River is given by $h_{y0}+z_y=H_{nc}$, where h_{y0}: initial flow depth of the Yakuin-shinkawa River, z_y: bed elevation of the Yakuin-shinkawa River, and H_{nc}: water level of the Naka River at the confluence of the Yakuin-shinkawa and the Naka River.

4.6 Flood flow in the Hakata River

Basic equations are eqns. (1) to (5). As the boundary conditions, flow depth and discharge can be given at the upstream and downstream end of the Hakata River in the same way as that described in the former sections.

4.7 Conditions for the simulation

Under the conditions shown in table 1, we compute the flood flow discharge and depth in the Naka, Yakuin-shinkawa and Hakata River.

5 Simulation results of flood flow in the Naka and the Yakuin-shinkawa River

The calculated floodwater level is shown in figs. 3 and 4. Corresponding to the beginning of heavy rainfall at 7:40, the Naka and the Yakuin-shinkawa River increases in water level.

Table 1: Simulation conditions for flood flow in the rivers.

Calculation period	0:00 to 24:00, June 29, 1999
Calculation distance (The Naka River) (from Minami-Ohashi bridge to the river mouth)	3.7km
Time step	$\Delta t =$ 1 sec
Distance step	$\Delta x =$ 20m
Manning's roughness coefficient	$n =$ 0.03
Correction factor related to energy loss	$\xi_c =$ 1.0
Runoff coefficient of the Yakuin-shinkawa River watershed	$f =$ 0.7

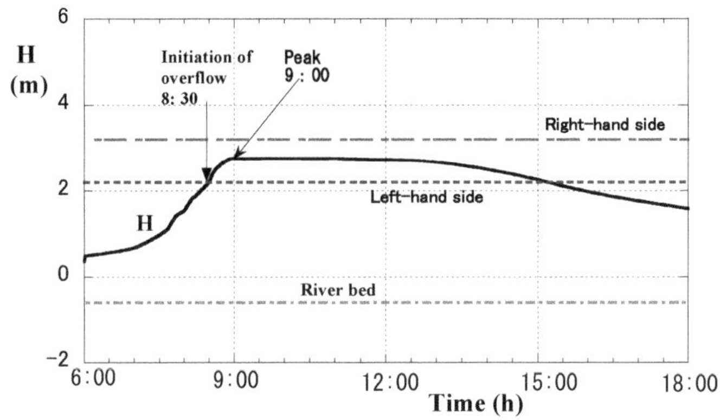

Figure 3: Time-varying water level at Haruyoshi bridge of the Naka River.

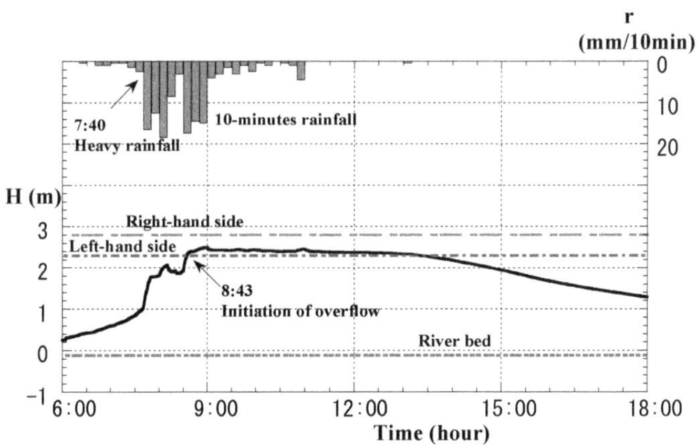

Figure 4: Time-varying water level at Shinkawa bridge of the Yakuin-shinkawa River.

The Naka River begins to overflow near Haruyoshi bridge at 8:30. The peak of overflow depth is at 9:00. The overflow lasts until 15:11. The Naka River also begins to overflow near Sumiyoshi Bridge at 8:45, and end the overflow at around 13:30. Furthermore, the Naka River overflows at the separation point of the Hakata River at 8:40, and end the overflow at 13:59.

The Yakuin-shinkawa River, on the other hand, increases rapidly in water level near Shinkawa bridge at 7:40. This river begins to overflow near Shinkawa bridge at 8:43. The overflow lasts until 12:08. The vicinity of the Shinkawa bridge has some entrances to the subway station called 'Tenjinminami-eki' and the underground shopping mall called 'Tenjin-chikagai'. Therefore, we can know the possibility of the inundation of the subway station and the underground shopping mall.

It takes only about one hour from the beginning of heavy rainfall to the initiation of the overflow. Therefore, we can see that it is important for the inhabitants and office workers in the downtown 'Tenjin district' to watch both the Naka and the Yakuin-shinkawa River during heavy storm events.

6 Two-dimensional urban flood simulation model [5, 6]

From Section 5, it is found that the Naka and the Yakuin-shinkawa River overflow their banks during heavy storm events; possibly floodwater from the Yakuin-shinkawa River moves down roads to the entrances of the subway station and the underground shopping mall. This indicates the possibility of inundation of the underground spaces. Section 6 describes a two-dimensional flood simulation model for this urban area. This simulation is composed of two steps; the first one is flood simulation on the ground level and the second one is on the subway station concourse (the 1st basement).

6.1 Basic equations

Basic equations are momentum and continuity equations for flood flows on a dense urban area [5, 6]. Two kinds of resistance terms are considered in the momentum equations; one is due to the shear stresses on the bottom and the other due to drag forces on the houses and buildings.

6.2 Initial and boundary conditions for flood simulation on the ground level

The initial bottom condition of inundation area is dry. As the boundary condition, overflow discharge q_{out} can be given at the overflow points on riverside by eqn. (4). We use the measurement of 10-minutes rainfall as rainfall intensity $r(t)$ from 0:00 to 24:00 on June 29, 1999. The drainage discharge q_{sink} per unit area can be estimated from the record of pump drainage at the stations in the inundation area during the 1999 Fukuoka flood event. The result is

$$q_{sink}=0.7\times 52.0 \quad \text{(mm/h)} \tag{9}$$

where the value of 52.0 (mm/h) corresponds to the capacity of the pump stations.

6.3 Initial and boundary conditions for flood simulation on the subway station concourse (the 1st basement)

The simulation estimates floodwater level at every calculation mesh on the ground level and at the entrances to the underground spaces. Since there are steps of steep slope at the entrances, hydraulically critical flow depth occurs at the entrances and determines inflow discharge of the underground spaces. This becomes the boundary condition for the flood simulation on the subway station concourse (the 1st basement).

6.4 Simulation conditions

Tables 2 and 3 show the conditions for the flood simulation on the ground level and on the subway station concourse (the 1st basement), respectively.

Table 2: Conditions for the flood simulation on the ground level.

Calculation period	0:00 to 24:00 on June 29, 1999.
Calculation area	6,000m×5,000m
Time step	$\Delta t = 1$ sec
Distance step (mesh size)	$\Delta x = \Delta y = 100$m.
Manning's roughness coefficient	$n = 0.043$
Drag coefficient C_D	$C_D = 2.05$
$q_{sink}=0.7\times52.0$ (mm/h)	

Table 3: Conditions for the flood simulation on the subway station concourse (the 1st basement).

Calculation period	7:30 to 17:30 on June 29, 1999.
Calculation area	60m×180m
Time step	$\Delta t = 0.005$ sec
Distance step (mesh size)	$\Delta x = \Delta y = 1$m
Manning's roughness coefficient	$n = 0.013$ for the passage
Manning's roughness coefficient	$n = 0.05$ for stairs
$q_{sink}=0.0$ (mm/h)	

7 Results and discussion of flood flow simulation in the downtown area

7.1 Inundation on the ground level

Figure 5 shows inundation depth and flux vector estimated at 9:00 from the two-dimensional flood simulation on the ground level.

The heavy rainfall from 7:40 results in inundation due to rain water on the downtown at around 8:00. For example, Imaizumi district becomes inundated because of its lower ground level. Afterwards the Yakuin-shinkawa River overflows to the left-hand side near Shinkawa bridge at 8:43. Floodwater spreads to the entrances of the subway station and the shopping mall. The peak of inundation occurs at 9:00 at these entrances (fig. 6). At this time, these entrances have inundation depth of 39cm–49cm. The water stop of the entrance to the subway station is 3.19m high (80cm deep). This is provided to prevent floodwater from entering the underground space. We can see that the water stop installed at the entrances is effective for preventing inundation of the underground space. However, it takes only about one hour and twenty minutes from the beginning of heavy rainfall to the peak of the inundation at the

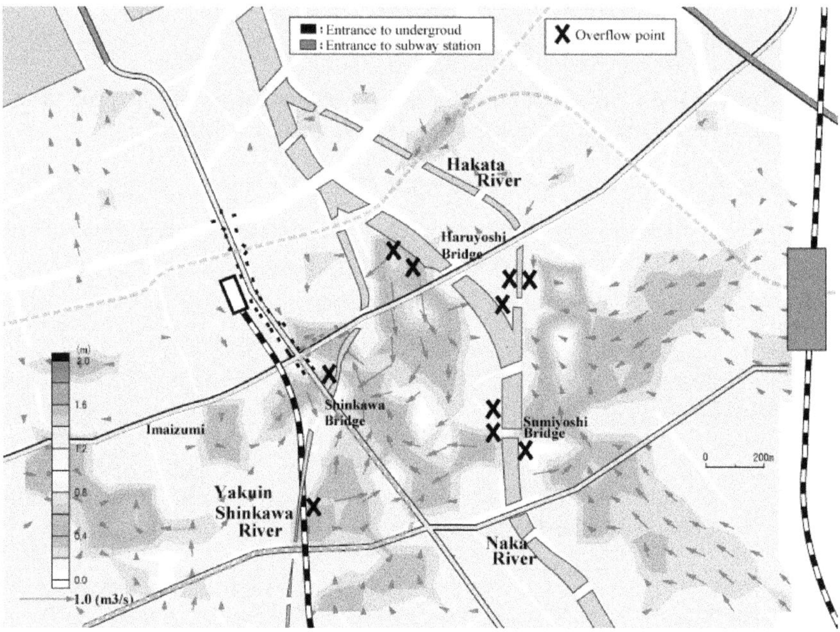

Figure 5: Inundation depth and flood flow velocity vector in the downtown at 9:00.

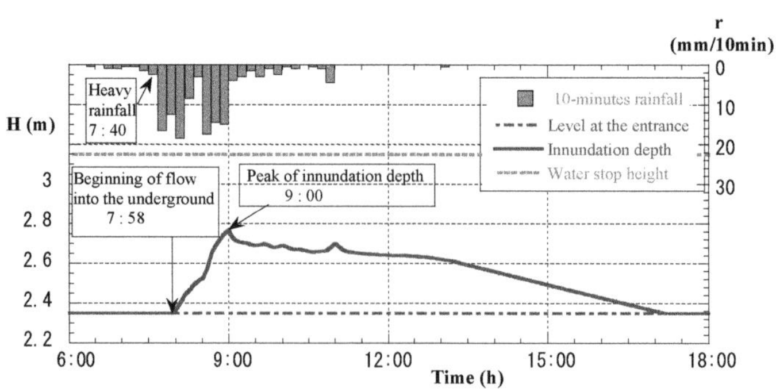

Figure 6: Time-varying inundation depth at the entrance to the subway station.

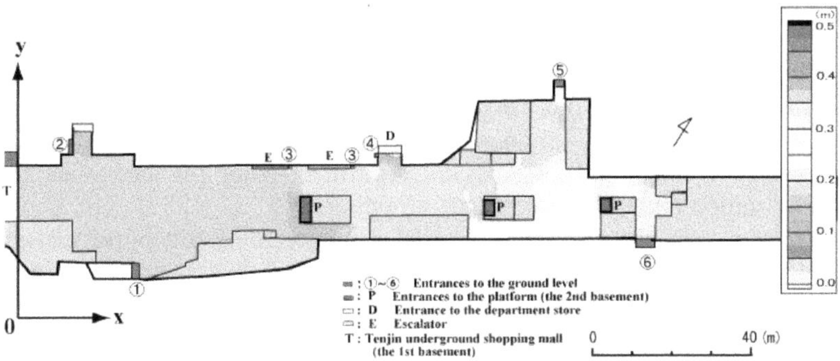

Figure 7: Inundation depth of the subway station concourse at 9:00.

entrances. It is possible for the administrators of the underground space to miss installation of the water stop before floodwater reaches the entrances. Therefore, inundation of the underground space is possible. In particular, inundation becomes deeper at the entrances 5 and 6 (fig. 7).

7.2 Inundation on the subway station concourse (the 1st basement)

Figure 7 depicts inundation depth at 9:00 determined by the flood simulation on the subway station concourse (the 1st basement). Floodwater begins to flow from the entrance 6 to the concourse at 7:58 and from the concourse to the platform (the 2nd basement) at 8:02. The whole concourse becomes inundated at 8:25 and the peak is at 9:01. The peak flow depth is 48cm.

Table 4 shows the amount of flooding water on each area of the subway station.

Table 4: Amount of floodwater on each area in the subway station and its vicinity.

Platform	23.8×10^3 m^3
Underground shopping mall	2.9×10^3 m^3
Shop	8.9×10^3 m^3

67% of inflow water moves down to the platform and 8% to the underground shopping mall (the 1st basement). We can see that the platform plays a role of drainage system for the underground space.

8 Conclusions

We have simulated the flood flow behaviour under the same rainfall condition as that on June 29, 1999. We have assessed flood risk of the Naka River watershed with the flood simulation.

The results obtained in this study are as follows:

1. The urban flood simulation shows that the heavy rainfall cause two different floods; one is due to overflowing water from the Naka and the Yakuin-shinkawa River and the other is due to the rain water on the downtown area.
2. The downtown called 'Tenjin district' is inundated by their superimposed floods; the result yields inflow of floodwater from the entrances to the subway station and the underground shopping mall.
3. Floodwater from the entrances 5 and 6 to the station concourse causes the inundation of the shopping mall (the 1st basement) and the platform (the 2nd basement). 67% of floodwater moves into the platform and 8% into the shopping mall. It is found that the platform plays a role of drainage system for the underground space.

Acknowledgements

This research has been supported by Foundation of River and Basin Integrated Communications, Japan and Grant-in Aid for Scientific Research, Japan Society for the Promotion of Science. The cross-sections of each river have been supplied by Fukuoka City Government. The result of runoff analysis at the Minami-Ohashi bridge has been given by Kyushu Branch, CTI Engineering Co., Ltd. The authors would like to appreciate their grant in aid and supply of the river data on this research.

References

[1] Hashimoto, H., Nonaka, Y. and Saito, M., Flood-risk Assessment of the Naka River in Fukuoka City, Japan, *Proceedings of the 8th International Conference on Urban Drainage Modelling,* Tokyo, Japan, CD-Rom, September, 2009.

[2] Fukuoka, S., Kawashima, M., Matsunaga, N. and Maeuchi, H., Flooding water over a crowded urban district, *Journal of Hydraulic, Coastal and Environmental Engineering,* JSCE, No.491/II-27, pp. 51-60, 1994 (in Japanese).

[3] Toda, K., Inoue, K., Murase, S. and Ichigawa, Y., Modeling of overland flood flow due to heavy rainfall in urban area, *Annuals, Disaster Prevention Research Institute, Kyoto University,* No.42, B-2, pp. 355-367, 1999 (in Japanese).

[4] Haider, S., Paquier, A., Morel, R. and Champagne, J.-Y., Urban flood modelling using computational fluid dynamics, *Water & Maritime Engineering,* 156, pp. 129-135, 2003.

[5] Hashimoto, H., Park, K. and Watanabe, M., Overland flood flow around the JR Hakata-eki station from the Mikasa and Sanno-Channel river in Fukuoka city on June 29, 1999, *Journal of Japan Society for Natural Disaster Science.* 21-4, pp. 369-384, 2003 (in Japanese).

[6] Hashimoto, H. and Park, K., Two-dimensional urban flood simulation: Fukuoka flood disaster in 1999, *Proceedings of the First International Conference on Flood Recovery, Innovation and Response,* pp. 59-67, 2008.

Damage and loss prediction model considering inundation level, flow velocity and vulnerability of building types

H. Maiwald & J. Schwarz
Earthquake Damage Analysis Center (EDAC),
Bauhaus Universität Weimar, Germany

Abstract

Reliable prognoses of building damage caused by flood impact require realistic relationships between action and damage or loss describing parameters. Due to the fact that commonly applied damage functions are related to the different usage classes (i.e. private housing), the required differentiation according to the parameters on the resistance side is still missing and further impact parameters like flow velocity remain unconsidered. On the basis of the Earthquake Damage Analysis Center (EDAC) flood database (including different datasets collected after the 2002 flood in Saxony), a method to determine the structural damage and the losses for any given flood scenario was developed. The key elements of the developed method are the engineered evaluation and categorization of the existing building stock, the definition of five damage grades with respect to the observed damage patterns and the classification of building types into vulnerability classes (and the consideration of the inherent scatter). Specific Functions enabling the calculation of damage grades and the loss assessment in separate steps of the same procedure. Specific Vulnerability Functions (SVF) are related to the inundation level or additionally the flow velocity (action side), and the predominant building types or the vulnerability classes (resistance side). Specific Damage Functions (SDF) can calculate the expected losses directly under consideration of the inundation level and the predominant building types (SDF of Type 1a) or the assigned vulnerability classes (SDF of Type 1b). SDF of Type 2 transfers the damage grades into loss statements. The developed tools are applied to the reinterpretation of the August 2002 flood damage. Results are presented for different study areas on a micro-scale level. In all cases, remarkably good agreement between the predicted and the reported losses can be

WIT Transactions on Ecology and The Environment, Vol 159, © 2012 WIT Press
www.witpress.com, ISSN 1743-3541 (on-line)
doi:10.2495/FRIAR120051

stated. First investigations with respect to the location of buildings, the direction of flow and the structural damage are also represented.
Keywords: damage grades, flood damage, damage functions, flow velocity, loss estimation, specific energy height, vulnerability classes, vulnerability functions.

1 Introduction

Estimating damage caused by flood impact is an important, yet scientifically and methodically insufficiently investigated task. The large scatter within the scarcely available damage data as well as the remarkable diversity (in shape and amplitude characteristics) of the statistically derived damage functions contribute to uncertain and often misleading loss predictions. The situation is still more complicated as harmonized damage assessment procedures and documentation manuals are missing.

This paper refers to the development of a damage and loss prediction model based on an engineering evaluation system of buildings [1, 2]. Alluding to the procedure developed in the risk analysis of earthquakes, it is checked, whether methodical fundamentals can be transferred or have to be adopted, and which parameters must be derived from data surveys. As an essential improvement, in addition to the inundation level, the flow velocity on the action side [3, 4] and the impact of structural parameters on the resistance side [5, 6] are considered.

Preliminary results of the approach are concentrated on the case studies of the towns Eilenburg, Döbeln and Grimma, being particularly affected by the 2002 flood in Saxony (Eastern Germany). Innovative damage and loss assessment procedures are developed including the unified definition of global structural Damage Grades (D_i) and Specific Vulnerability Functions (SVF), Specific Damage Functions (SDF) being related to the damage and loss or the specific flood vulnerability classes (HW-VC) of a building or object. As is it can be shown, the tools enable the reinterpretation of damage with respect to structural and loss parameters, and are, therefore, suited for prognoses purposes and short- and long term disaster management decisions.

2 Data base

The key element of the procedure lies in the preparation of the real damage cases, which were elaborated immediately after the 2002 floods in Saxony.

- Dataset 1 is related to the outcome of a questionnaire survey performed by the authors between 2002 and 2004 [5].
- Dataset 2 is the outcome of telephone calls after the 2005 Danube flood in Bavaria and 2006 Elbe flood in Saxony (performed within the frame of the RIMAX project MEDIS [7]).
- Dataset 3 is the outcome of an evaluation of damage expertises by order of the “Landestalsperrenverwaltung Sachsen (LTV)” [8].

After having unified the data with respect to the structural parameters and having transferred the verbal damage descriptions into the proposed scheme of damage grades, the datasets could be considered separately and as a combination. The datasets include information about duration, velocity (qualitatively) and other secondary (probably damage contributing) flood action as well as vulnerability-related parameters. Dataset 1 and 2 are quite complementary with respect to parameter ranges of inherent data points, and lead to a well distributed database. Dataset 3 is the outcome of an evaluation of qualified damage expertises and is therefore considered separately in the following investigations.

Nevertheless, due to the differences within, the derived vulnerability functions have an impact on the prognosis results (see Figure 1).

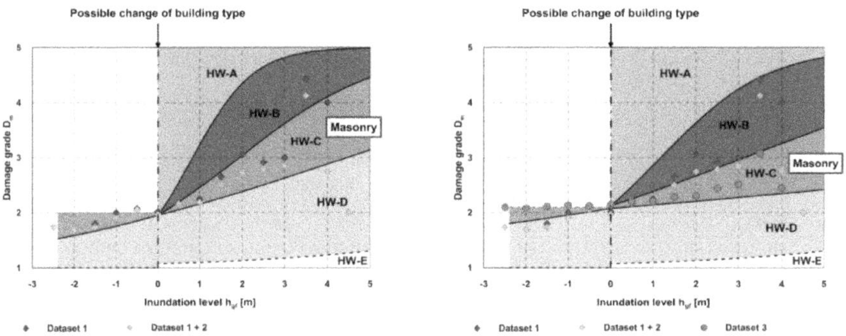

a) State 2008 [4] b) State 2012

Figure 1: Damage grades, inundation level and vulnerability classes; results for masonry buildings.

3 Basic elements of the flood damage and loss prediction model

During the last years, and as outcomes of practical requests, several research projects of the EDAC have been concentrated on the development of an engineering evaluation system of buildings subjected to natural hazards and the elaboration of more refined tools to link elements of hazard, action, vulnerability, damage and loss in a modular way [1, 2, 4]. The procedures and the processing levels implemented in the model are structured transparently and can be used for different risk types (earthquake, storm, flood, tsunami etc.).

Basics steps of the procedure are derived from analogous considerations to the empirical, intensity-oriented method introduced for the earthquake damage and loss model on the basis of EMS-98 [9].

Mainly focusing the consideration on structural damage due to flood impact, characteristic vulnerability classes are determined for the different building types. Their vulnerability functions are derived by the following step-by-step procedure.

3.1 Damage descriptions and assignment of repeatedly observed effects

Field surveys have to be qualified with respect to the documentation of damage cases and the collection of structural parameters affecting the vulnerability of each building. The documentation has to implement an "engineered" description of the building and its structural damage, and to archive typical damage patterns. Repeatedly observed effects are used as indicators for the definition of damage grades. In addition to the structural damage, observed damage-indicating phenomena can be related to chemical or physical origin [6]. For the damage classification, rehabilitation measures are of importance to convert the visible action or technological term into a generalized scheme of damage interpretation. Table 1 summarizes the main criteria for the classification of observed effects and damage reports, in an extended and widely applicable format.

Table 1: Assignment of damage grades D_i to damage cases [4, 6].

D_i	Damage		Description	Drawing	Example
	Structural	Non-structural			
D1	no	slight	only penetration and pollution		
D2	no to slight	moderate	slight cracks in supporting elements impressed doors and windows contamination replacement of extension elements		
D3	moderate	heavy	major cracks and / or deformations in supporting walls and slabs settlements replacement of non supporting elements		
D4	heavy	very heavy	structural collapse of supporting walls, slabs replacement of supporting elements		
D5	very heavy	very heavy	collapse of the building or of major parts of the building demolition of building required		

3.1 Definition of damage grades

Repeatedly observed effects can be regarded as typical building response indicators for a comparable level of damage, loss of integrity, stability etc.

Table 1 provides the background for the necessary generalization of any damage classification. By the definition of damage grades (D_i), a unified evaluation of all damage data and reports is guaranteed. Damage grades enable the logical link between flood impact and loss in an innovative way. In all cases a minimum damage grade D1 (without the occurrence of structural damage) has to be assigned due to humidity penetration effects. The generalized damage definitions are related to the quality of structural damage and non-structural damage as well as to the required extent of rehabilitation or other damage replacement measures (cf. Table 1).

3.2 Correlation between flood impact parameters and building damage (D_i)

In the majority of cases, the damage descriptions submitted by the questionnaires (dataset 1), telephone calls (dataset 2) or the expertises (dataset 3) have to be translated into damage grades. This essential work has been performed by engineers from EDAC staff on the basis of the developed evaluation tools (Table 1). For each damage case, the damage grade D_i ($i = 1$ to 5) and the flood action parameters could be assigned. On the basis of sample individual damage grades (i.e. for the same building type under comparable inundation heights), the mean damage grade (D_m) can be determined, which is a robust parameter for damage prognosis (see Figure 1).

The level of the ground floor is taken as a reference height for the flood impact (h_{gf}). By this definition of inundation height, it is recognized that at this level an abrupt change of building material and/or structural systems (basement to ground floor) and, consequently, of vulnerability might occur. Therefore, in the vicinity of $h_{gf} = 0$, discontinuities in the vulnerability function curve indicate the new quality of the approach, i.e., vulnerability classes of basement and other building floors have to be assigned separately (see Figure 1).

For the all masonry buildings included in the datasets, mean damage grades D_m are calculated for intervals of inundation height (Δh). Results of this procedure are illustrated for the predominant building type (masonry wall structures). Taking for the interval of inundation height $\Delta h = 0.5$ m, the increase of mean damage grade (D_m) with the impact parameter becomes evident. The dots derived from the basic datasets of this study create the new type of damage functions. Figure 1a shows the outcome of the investigations in [4, 6]. As a result of the consideration of the more qualified dataset 3, the borderlines for the vulnerability classes have to be adjusted (Figure 1b). Further investigations are necessary.

3.3 Ranges of vulnerability classes for the predominant building types

For each building, the characteristic building type (or structural system) and its vulnerability class have to be identified. "Vulnerability" is taken as a measure for the resistance of a building against comparable impact conditions (inundation height, flow velocity) and is related to the differences in the damage (or loss) under these action parameters. Vulnerability functions are expressions of the existing correlations. Their quality can be improved by assigning the typical ranges and the scatter for the regionally predominant building types. Buildings of different structural type and material belong to the same vulnerability class, if for the relevant range of flood action parameter, similar mean damage grades have to be expected (Table 1). Due to the uniform quality of the database, it was possible to identify typical shapes of (still idealized) vulnerability describing functions (see Figure 1).

As a whole, five Flood Vulnerability Classes (here: HW-A to HW-E) are distinguished by definition, covering the range from low flood resistance/higher vulnerability (A – very sensitive; B – sensitive), to normal (C) and increased flood resistance (D).

Hypothetically, a flood resistant design (FRD) would lead to class (HW-E). Class HW-E buildings (as recommended in common guidelines) are characterized by a separation of building from the flood water table, for instance, by "up-lifting" the base floor over a grid of story-high columns. To convert each building of the dataset into its vulnerability class, a classification scheme is required, which takes into account the data density as well as the scatter within the representative samples of each building type (see Table 2).

For each building the characteristic structural system (here denoted as building type) has to be identified. Subsequently, the building types have to be sorted into their appropriate vulnerability classes, where most likely, still probable and also exceptional cases have to be considered (Table 2). The symbols in Table 2 replace empirical vulnerability functions in a robust way. The lines (full, broken) indicate the range of scatter and the probability of occurrence.

The vulnerability of building types can cover ranges of two or three vulnerability classes. If the user of the scheme is untrained in engineering practice or cannot decide about the extent of vulnerability affecting particularities (due the lack of information), the most likely vulnerability has to be taken.

3.4 Specific Vulnerability Functions (SVF)

3.4.1 Correlation between damage grade D_i and inundation level h_{gf}

The innovative elements of the whole procedure can be subsumed and linked by a new type of (SVF), systematically developed and continuously presented depending on the progress of data elaboration [5, 6, 10]. The database enables the differentiation of these functions with respect to the main structural (wall) material (SVF type 1a) or, alternatively, with respect to the flood vulnerability class (SVF type 1b). Examples for the latter, reflecting the outcome of the combined dataset 1+2, can be taken from Figure 2a [2]. Figure 2b shows the results derived from the more comprehensive dataset 3.

Table 2: Classification of building types in vulnerability classes and identification of ranges of scatter (on the basis of the evaluated data).

Classification of building type			Flood vulnerability class HW-VC				
Main building type	*short*	Example	A	B	C	D	E
Clay	*Clay*		O				
Prefabricated	*PF*		|—	—O—	—|		
Framework	*FW*		|—	—O···	···|		
Masonry	*MW*		|···	···—	—O···	···|	
Reinforced concrete	*RC*				|···	···O	
Flood resistant design	*FRD*					|—	—O

O Most likely vulnerability class
— Probable range
▪▪▪ Range of less probable, exceptional cases

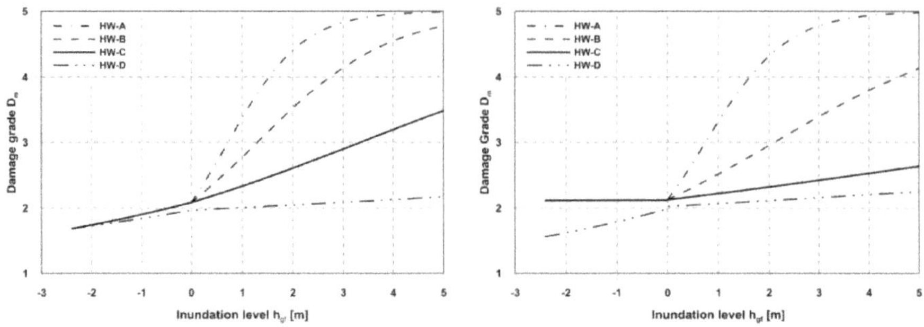

a) SVF type 1b: Dataset 1 + 2 b) SVF type 1b: Dataset 3

Figure 2: Specific vulnerability functions of type: $D_m = f\,(h_{gf})$.

3.4.2 Correlation between damage grade D_i and the specific energy height

A first outcome within the framework of the RIMAX-project “MEDIS” for consideration of flow velocity was presented in [4].

For a few of the flooded areas in the 2002 event in Saxony, information about the recalculated flow velocity (v_{fl}) was submitted by the Regulatory Office (LTV).

It has to be highlighted that for a rather limited number of damage cases the relevance of flood velocity (v_{fl}) in combination with the inundation level above ground level (h_{gl}) could be investigated. The first approach related to flood intensity ($h_{gl} \times v_{fl}$) failed because no clear tendency could be established [1]. In a second approach, the specific energy height $H = h_{gl} + (v_{fl}^2/2g)$ is predicted. Due to the limited number of samples, the building stock remains undifferentiated (being representative for masonry buildings). As a result of this first data check, a building material independent vulnerability function (SVF type 2*) is given by Figure 3a.

It can be concluded that mean damage grade D_m remains nearly constant for $H < 2.0$ m; for $H > 2.0$ m, a steady increase of damage grade can be derived from the observations, which is mainly attributed to the impact of increasing flow velocity (v_{fl}). From an engineering point of view, this clear tendency seems to be explainable.

Further investigations in the same flooded areas, but with a larger number of damage cases (taken from dataset 3), are published in [3]. Figure 3b shows preliminary vulnerability functions for vulnerability classes (SVF type 2b). Due to the still relatively low data volume for the vulnerability class HW-A, the corresponding vulnerability function is still an engineering-based assumption.

The starting values for the damage grades D_m at $H = 0$ m results from flooding in the basement area, because about 80% of the examined damage data regarding structures with basements.

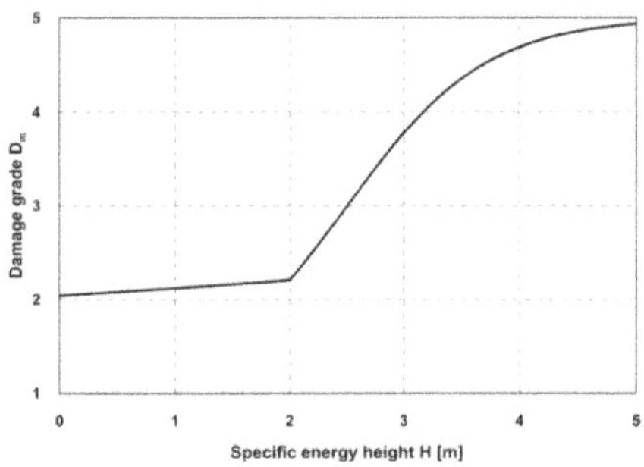

a) SVF type 2*: Dataset 1 [4]

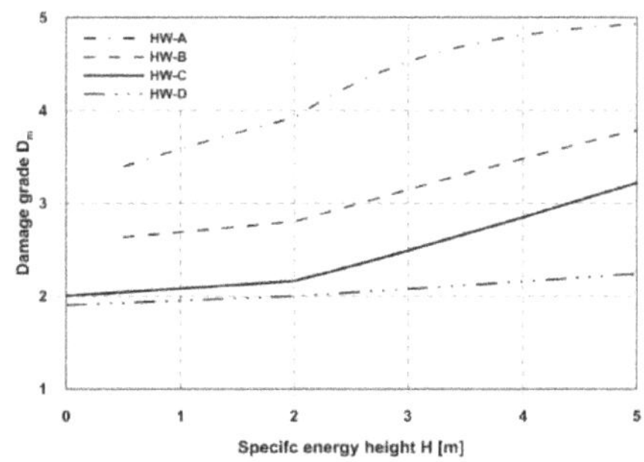

b) SVF type 2b: Dataset 3 [3]

Figure 3: Specific vulnerability functions of type: $D_m = f(h_{gl}, v_{fl})$.

The vulnerability classes HW-A and HW-B are generally not found in the basement area. Therefore, the corresponding vulnerability functions in Figure 3b are shown only for a specific energy height level of $H > 0.5$ m. The characterization of the qualitative features of the functions for $H < 0.5$ m remains to be determined by further investigations on a larger dataset. Further research is required to establish damage prediction models in zones where a sudden increase of flow velocity is expected due to the hydrological and topographical situation.

The parameters for the different types of vulnerability functions (with exception of the functions in Figure 2b) could be taken from [2] .

3.5 Location Specific Vulnerability Functions (LSVF)

The location of the structure relative to the flow velocity vector (flow direction) and the urban-structural integration of the structure have an influence on the structural damage. Table 3 shows a preliminary proposal for the classification of the relevant situations from an engineering point of view on the classification of a building into the surrounding housing and the flow characteristic during flood events [3]. Real structures of settlement are usually complex; therefore, a simplified classification scheme does not cover all situations. For example, if the flow directions change during a flood event. Buildings are also often not orthogonal and tangential to the incident flow direction, so that a classification according to Table 3 is clearly not possible. In addition, the flow direction can be set for complex site conditions only by specifying a velocity vector. The classification scheme in Table 3 is therefore to be understood as a first proposal and has to be revised on a larger database.

Figure 5 shows an attempt to classify different situations from the real development study areas, assuming a main flow direction according to the scheme in Table 3. Analogous to the procedure in 3.3, the mean damage grades D_m are calculated at intervals of $\Delta H = 0.5$ m.

Table 3: Classification scheme with respect to the location of buildings and flow direction (cf. [3])

Nr.	Type/Location	Description	Flow direction
1	Stand alone		Directly
2a	Front house	Beginning of a row of houses	Directly / flow around
2b	End house	End of a row of houses	Flow around / circulation
2c	Front/ End house	Beginning / end of a row of houses	Orthogonally/ circulation
3a	Central house	In the middle of a row of houses	Tangentially
3b	Central house	In the middle of a row of houses	Directly / orthogonally
4	Corner house	Cross situation	Flow around / circulation

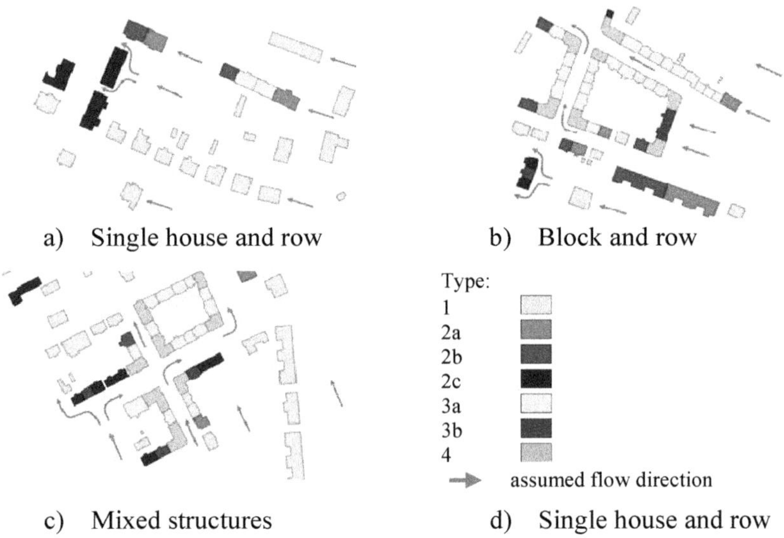

Figure 4: Examples for classification of real structure of settlement (Table 3).

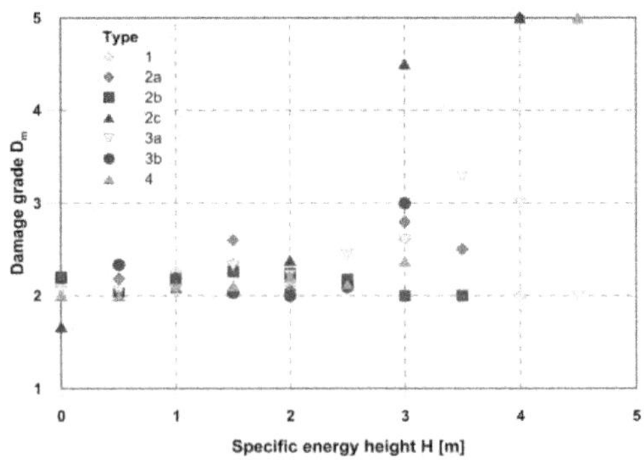

Figure 5: Mean damage D_m grades depending on specific energy height H, location of building and flow direction [3].

The results are shown in Figure 6. Due to the relatively small data density for several types according to Table 3, the results are not explicit in all intervals. However, it can be identified that the structural damage is higher for orthogonally streamed buildings (Type 1, 2a, 2c with the exception of Type 3b) than for tangential streamed buildings like Type 3a or Type 2b at the end of a row of houses. Further investigations on a more comprehensive dataset are necessary.

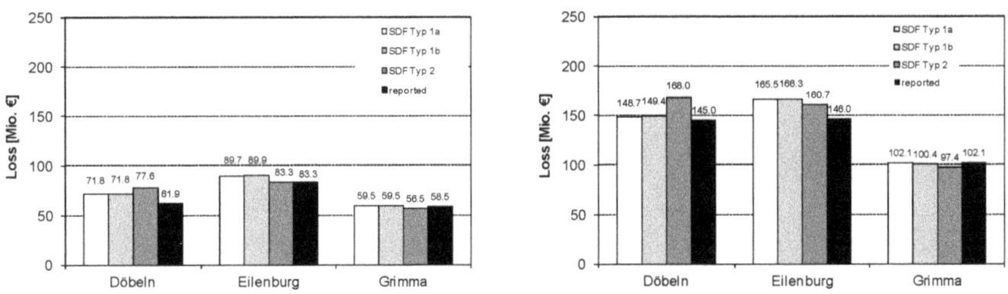

a) Residential buildings b) Entire building stock

Figure 6: Comparison between calculated and reported losses.

3.6 Specific damage functions (SDF) - loss prediction

Following the proposed methodology, a set of rather new types of Specific Damage Functions (SDF) for loss predictions were presented in [2, 3]. Functions refer to the building type (SDF Type 1a) or flood vulnerability class (SDF Type 1b). A second type of functions (SDF Type 2) transfers the calculated damage grades D_i (see Table 1) into loss statements.

All of the developed functions consider the number of stories and the presence of a basement [6]. The mathematical background of the different types of vulnerability functions is based on an exponential approach and is outlined in [2, 3, 10]. The parameters could be taken from [2].

4 Case studies

The innovative options and advantages of the whole procedure are demonstrated by the case studies from the 2002 flood in Saxony. With the developed specific vulnerability functions, the mean damage grade D_m can be calculated for each individual building (micro-scale level) in the three study areas Döbeln, Eilenburg and Grimma. In [2, 3], it can be shown how heavily the areas of a community will be affected and where significant damage concentrations are expected. With the different types of damage functions, loss statements are available for the study areas. For the re-interpretation with regard to the water level as the action side, damage functions SDF type 1a and 1b will be used.

In a third calculation variant, first the mean damage grades D_m with vulnerability functions SFV type 2b (with consideration of the flow velocity) are calculated.

These are then converted with damage functions SDF type 2 into loss statements. Despite the different flooding characteristics in the study areas, the comparison of calculation results with the official damage statistics of the Saxon Relief Bank (SAB) shows a very good agreement for the residential building stock (Figure 4a) for all the calculation variants. The slightly higher calculated losses in Döbeln are due to the fact that in the data accumulated in the SAB, not all damage cases are considered. For the prediction of the damage on the entire building stock (Figure 4b), it is taken into account that the application of the damage functions on non-residential buildings is only an approximation.

5 Photo credits

The pictures in Table 1 and 2 were made by the Earthquake Damage Analysis Center (EDAC) at the Bauhaus Universität Weimar.

References

[1] Earthquake Damage Analysis Center, EDAC-Flood Damage Model, http://www.edac.biz/research/floods/edac_hochwasserschadensmodell.html 2011.

[2] Maiwald, H., Schwarz, J., Ermittlung von Hochwasserschäden unter Berücksichtigung der Bauwerksverletzbarkeit. EDAC-Hochwasser-schadensmodell. scientific technical reports 01 (2011), Zentrum für die Ingenieuranalyse von Erdbebenschäden, Universitätsverlag, Bauhaus-Universität Weimar, 2011.

[3] Maiwald, H., Schwarz, J.: Berücksichtigung der Fließgeschwindigkeit bei Hochwasser-Schadensmodellen, Bautechnik 86 (2009) 9, 550 - 565, 2009.

[4] Schwarz, J., Maiwald, H.: Damage and loss prediction model based on the vulnerability of building types. Proceedings 4th International Symposium on Flood Defence, 6 – 8 May 2008, Toronto, Canada, 2008.

[5] Schwarz, J., Maiwald, H., Gerstberger, A., Quantifizierung der Schäden infolge Hochwassereinwirkung: Fallstudie Eilenburg. Bautechnik 82 (2005) 12, 845 – 856, 2005.

[6] Schwarz, J., Maiwald, H., Berücksichtigung struktureller Schäden unter Hochwassereinwirkung, Bautechnik 84 (2007) 7, 450 – 464, 2007.

[7] Schwarz, J., Maiwald, H., Von der Schadensaufnahme zur Verletzbarkeitsfunktion – ein Ansatz aus den Ingenieurwissenschaften. Abschnitt 9.2 in „Hochwasserschäden – Erfassung, Abschätzung und Vermeidung“. Abschlussbericht des RIMAX-Forschungsprojektes MEDIS: „Methoden zur Erfassung direkter und indirekter Hochwasserschäden”: (Hrsg. Thieken, A. H., Seifert, I., Merz, B.), oekom-Verlag, 2009.

[8] Schwarz, J., Maiwald, H., Ingenieurmäßigen Beschreibung der Schadenserwartung von Gebäuden unter Hochwassereinwirkung.

Abschlussbericht zum Projekt 246 143 75 im Auftrag der Landestalsperrenverwaltung des Freistaates Sachsen. Zentrum für die Ingenieuranalyse von Erdbebenschäden, Bauhaus-Universität Weimar, 2009.
[9] Grünthal, G. (ed.), Musson, R., Schwarz, J., Stucchi, M., European Macroseismic Scale 1998. Cahiers de Centre Européen de Géodynamique et de Seismologie, Volume 15, Luxembourg, 1998.
[10] Maiwald, H., Ingenieurmäßige Ermittlung von Hochwasserschadenspotentialen im mikroskaligen Maßstab. Dissertation, Department of Civil Engineering, Bauhaus-Universität Weimar, 2007.
[11] Kreibich, H., Piroth, K., Seifert, I., Maiwald, H., Kunert, U., Schwarz, J., Merz, B., Thieken, A. (2009) Is flow velocity a significant parameter in flood damage modelling? Natural Hazards and Earth System Sciences 5, 1679-1692, 2009.

On-line early warning system for evacuation of socially vulnerable population during flooding

D. Mioc[1], K. K. Moreiri[2], J. N. Nkhwanana[2], F. Anton[1], B. Nikerson[3], E. McGillivray[4], A. Morton[4] & P. Tang[5]
[1]*National Space Institute, Technical University of Denmark, Denmark*
[2]*Department of Geodesy and Geomatics Engineering, University of New Brunswick, Canada*
[3]*Faculty of Computer Science, University of New Brunswick, Canada*
[4]*New Brunswick Emergency Measures Organization, Canada*
[5]*New Brunswick Department of Environment, Canada*

Abstract

Natural disasters normally find people unprepared and emergency planners are faced with a big task of evacuating people. The response time needed to evacuate people, especially those that are socially vulnerable is very important in saving lives. There exist a number of important factors when planning for an evacuation, e.g. the number of people to be evacuated, time available for the evacuations, the distance to travel and also the available routes for evacuation. The recent flooding in Fredericton has identified that the socially vulnerable population require more resources and emergency planning than the evacuation of the rest of population. The socially vulnerable population needs to be identified before the disaster occurs and their special needs need to be documented. The provisions of medicaments, special food and any additional resources have to be planned and prepared in advance. This paper presents the approach to identify, map and assist the evacuation of the population that is socially vulnerable during floodings in Fredericton while taking care of their special needs. The main result of our research is a web based GIS system that provides appropriate information to the relevant authorities and general public in a timely manner and easy to understand.
Keywords: flooding, evacuation, early warning system, vulnerable population.

WIT Transactions on Ecology and The Environment, Vol 159, © 2012 WIT Press
www.witpress.com, ISSN 1743-3541 (on-line)
doi:10.2495/FRIAR120061

1 Introduction

The main goal of our research is to create a web based Geographic Information Systems (GIS) application, which would help people living in flood plains, who may at one point be vulnerable due to their proximity to the river and the amplitude of the flooding. There are many factors that may lead to floods, and people living in flood plains are most likely to be affected. A flood plain is whereby a flat or gently sloping land adjacent to a stream, during high stream flow conditions, caused by excessive rainfall or snowmelt or a combination of the two, becomes inundated by water [1]. A large number of people in the Fredericton area reside along the Saint John River and its tributaries. We wanted to create a web based GIS application [2] that would provide timely flood information to emergency managers and general population, and the main functionalities of this application are to help identify people within certain distances within the floodplain, that may be affected by a particular flood. Web based GIS [3] is a methodology of building distributed GIS applications on the internet [4]. Web GIS offers the possibility of querying on a website that can help to identify the properties and the people who might be affected by the floods and who may then be contacted to have ample evacuation time.

Natural disasters normally find people unprepared and emergency planners are faced with a big task of evacuating people. This paper investigates how mapping of social vulnerable population can assist emergency planners in successful evacuation and allocation of resources. According to Hansjurgens [5],

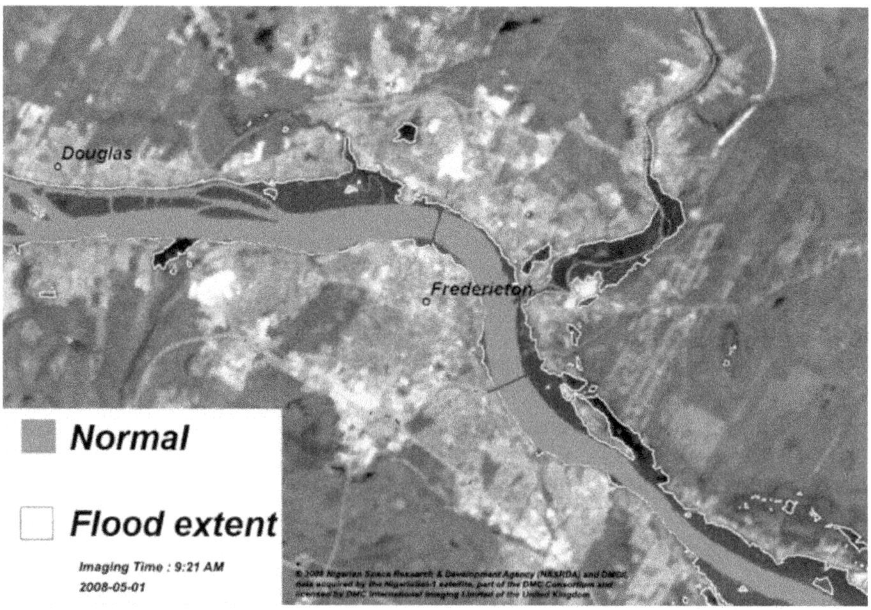

Figure 1: Satellite image of the study area taken during the flood in 2008.

social vulnerability refers to the characteristics of a person or group and their situation that influence their capacity to anticipate, cope with, resist or recover from the impact of a hazard. He also argues that people/societies may be vulnerable to floods in terms of being less likely to be aware of flood risk, secondly, being less able to cope with it and respond to the threat, and thirdly, being less able to recover afterwards. This project investigates the vulnerability and the results maybe used to help in evacuation and recovery.

Flooding is common in Fredericton and the 1973 flood (see Figure 1) was considered to be the worst according to the officials of City of Fredericton [6], and recently in 2008 CBC News [7] reported that about 40 streets were expected to be closed with water rising to levels above the 1973 flood. The same report mentions that 1300 homes would be affected in the process.

This paper also proposes that the mapping of this vulnerability should be web based [8]. This will allow emergency people to be able to access the information from the evacuation areas without having to rely on the GIS data sitting back at their offices. Watson [9] argues that it is important to get maps instantly with every team member. He also refers to the new technology that bridges the gap between a geospatial expert and a non-expert.

Usually, evacuation planning takes into account the following information:

- Characteristics of the hazard e.g. speed, duration, impact;
- Number of people to be evacuated;
- Time available;
- Distance to travel to ensure safety;
- Evacuation routes.

The evacuation process during flooding also requires information on accessible roads.

The vulnerability map will also show the location of temporary shelters to accommodate the population. It will also show the location of hospitals such that the emergency planners can easily locate the nearest hospital to the evacuation area when and if required.

2 Study area

It has been identified that more than 58% of the incorporated municipalities in the region of New Brunswick, reside within the Saint John River System [10]. The reason being that the waterways, during early settlement years, served not only as a source of major transportation for the early settlers, it also served as a source of livelihood for them as well. As the New Brunswick communities grew, so did the occupation of flood plains to the point where substantial damages, human sufferings and indeed, loss of lives occurred during major floods [1]. Furthermore, a publication by Jeanne Andrews [11] on floodings, highlighted the fact that flooding in Canada was a common phenomenon, resulting from an increase in stream flow beyond the point where the normal stream channel could contain the water. However, in most cases flood prone areas are often very attractive to the developer.

3 Prototype application design and implementation

The main goal of our research was to create of a web based GIS system which could provide appropriate information to the relevant authorities in a timely manner and easy to understand. Relevant authorities in this case may include flood managers, flood warning practitioners, professional partners and emergency services to those affected [12]. Furthermore, we wanted the flood warnings available online and readily accessible to a broad audience to increase the level and quality of services provided [13].

The easy to use interface (see Figure 2) should also allow non-GIS experts to be in a position to interactively view and explore, as well as query the database, to select different data variables and to view maps at several levels of detail. We used ArcIMS to develop the web based GIS system [2]. When creating this website, it was taken into consideration that the provision of a map based information system in an easily accessible manner presents many challenges and as such, most search results displayed on the map would also be accompanied by textual sections below the map.

Our other objectives included:

- To assist flood warning practitioners as well as the general public to be in a position to query the website for flood evacuation routes, as such the website would allow them to identify evacuation routes as well as to have an idea of the water levels on flooded roads (see Figure 2).

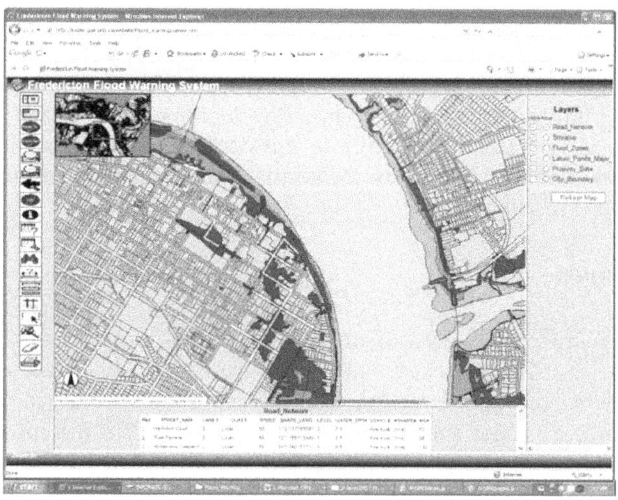

Figure 2: Output of a manually entered query.

- Design floods have also been utilised, for the system to have the capability of producing property data of properties that may be affected by a particular design flood. The design floods used are of 1 in 20 year flood, that is a flood having a return period of 20 years on average or having a 0.05% chance of occurring in a given year. Other design floods are 1 in 100 year and 1 in 200 year floods. These are intersected with the land use classification layer that includes industrial areas, offices, parks, shopping areas and residential blocks of people residing within the flood plain.

- This can be very helpful to flood practitioners to have an idea of the properties likely to be affected by such a flood (see Figure 4). Furthermore, city planners may also utilize the data when planning for future developments on flood prone areas to have an idea on the extent of the flood coverage of a possible maximal flood (shown on Figure 1).

Areas highlighted in yellow show the properties likely to be affected by a water level of three, which is up to 0.5 meters. In total, 243 properties would be affected by this water level rise (shown in Figure 3 and in Table 1 below):

Table 1: Identified property data.

240	2487100.8	7441993.95	01413871	100	1973	3	0	2	[polygon]	17486
241	2487144.05	7441998.15	01413384	100	1973	3	0	2	[polygon]	17564
242	2487176.95	7442030.8	01415140	0	1973	3	0	2	[polygon]	17618
243	2486935.05	7441933.5	75237180	100	1973	3	20	2	[polygon]	17787

Previous 25 Records   Zoom to these records

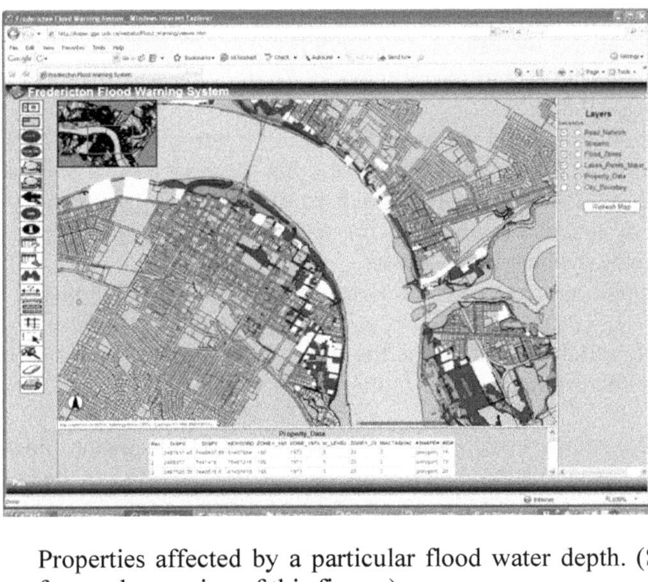

Figure 3: Properties affected by a particular flood water depth. (See online for a color version of this figure.)

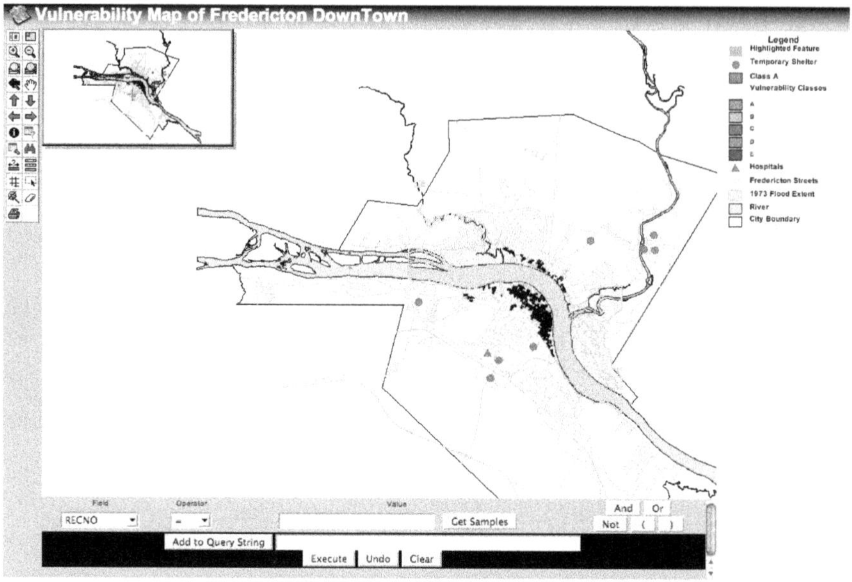

Figure 4: Web GIS application for mapping the socially vulnerable population during flooding.

4 Mapping of social vulnerability

Now we will present the mapping of the social vulnerability of each parcel that was affected by the 1973 flood. The 1973 flood extent was chosen because it was one of the worst floods in recent times in Fredericton and also because the data was readily available.

The mapping of social vulnerability was first introduced by [14]. There exists some calculation for social vulnerability, but they have all looked at blocks of population and not necessarily at each parcel [14, 15]. Chakraborty *et al.* [15] introduced a social vulnerability for Evacuation Assistance Index (SVEAI), which combines vulnerability factors.

This formula was used for the Hillsborough County, Florida and it works as follows:

1. For each variable *i*, determine the ratio of the variable in the block group to the total number of that variable in the county (*Ri*).

2. Compute a standardized Social Vulnerability for EvAcuation Index (SVEAI*i*) for variable *i* using the maximum ratio value *R*max observed in the county.

$$\text{SVEAI}i = \frac{Ri}{R\max}$$

3. To combine multiple variables in the assessment of social vulnerability, calculate the arithmetic mean of the vulnerability indices by dividing the sum of index values of all variables by the number of variables (n) considered.

$$\mathrm{SVEAI} = \frac{\sum \mathrm{SVEAI}i}{\mathrm{N}}$$

This formula was adopted from Cutter *et al.* [6] who introduced a Social Vulnerability Index (SOVI). Cutter's formula is used to calculate social vulnerability for counties.

The above-mentioned approaches work with groups of people, but this paper attempts to demonstrate that mapping information at parcel level can be more useful to emergency planners during evacuation.

Cutter *et al.* [14] identified the following as indicators of Social Vulnerability:

- Age, gender, race, and socioeconomic status;
- Characteristics depicting special needs populations or those that lack the normal social safety nets necessary in disaster recovery;
- The quality of human settlements (housing type and construction, infrastructure, and lifelines) and the built environment.

The approach of this paper will not be based on a formula or any weighting system. The indicators will be mapped as they are or grouped according to level of assistance required.

Based on the past experience of emergency managers in New Brunswick, the following classification is used in our application:

The first class (see Table 2) will comprise households that have people requiring medical assistance or population aged over 85 years. Then, we will have another group of people above the age of 5 and who have disabilities.

The third class will be people with kids under the age of 5 and the fourth class will be families without cars or living in mobile homes.

The last class will be people who can evacuate with little or no outside help. The table below summarizes the classes we are using in our application.

Table 2: Vulnerability classes.

Vulnerability Class	Description
A	Population requiring medical assistance or Population above the age of 85 years
B	Population with disabilities
C	Kids under the age of 5
D	Families in mobile home or families without cars
E	Population capable of self evacuation

Given this kind of information, emergency planners can know which households to assist first and also what resources to carry during the evacuation process. It is also acknowledged that a household can comprise one or more of the above classes, but in this case a household is given a class of the person/persons requiring the most assistance.

5 The web GIS application for evacuation of socially vulnerable population

For web based implementation of the early warning system for evacuation of socially vulnerable population we decided to use ESRI ArcIMS software.

ArcIMS offers two main templates, one based on HTML and another that runs through Java applets. The HTML template and the ArcGIS server have an advantage in that they do not require installation of Java on a client's computer to run. On the other hand a Java template offers an editing capability in the form of map notes that can be very useful for emergency planners.

Given the MapNotes functionality, users can create notes of their application and share with other users of the same Internet GIS application.

To give an example, each team can easily create a new point symbol for each household they have evacuated. This information can then be shared with other teams so that teams that have finished their work can know which households still require assistance. At the same time the progress can be centrally monitored to ensure that people are evacuated timely or if there is extra help required.

The ArcIMS HTML view was adopted for this paper. The application has eight datasets shown on Figure 5:

- Vulnerability classes;
- Class A (from vulnerability classes);
- Temporary shelter;
- Hospitals;
- 1973 Flood extent;
- St. Johns river;
- Fredericton streets;
- City boundary.

The database for Class A was created separately to show medical conditions of the people requiring assistance during evacuation. Given these extra details, the emergency planners can therefore select the appropriate shelters for the evacuated population using information from the shelters (see Table 3). The table below summarizes the queries available [8].

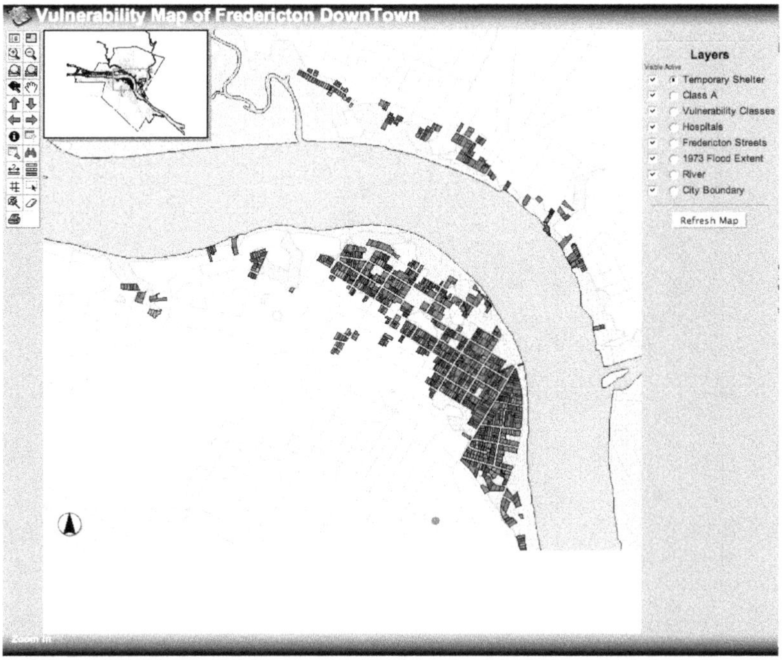

Figure 5: Data layers on the map available for the queries.

Table 3: Summary of the queries available within the application.

Dataset	Queries available
Vulnerability Classes	Number of Residents, Vulnerability Class
Class A	Number of Residents, Medical Condition
Temporary Shelter	Shelter Name, Capacity, Available support equipment
Fredericton Streets	Street Name

In our application we provided sample values for each query for non-experienced users (see Figure 6).

ArcIMS allows queries of one dataset at a time, i.e. a user cannot query two or more datasets at the same time. For example, a user can only query Vulnerability classes and it cannot combine the query with one from Temporary Shelters (shown in Figure 7).

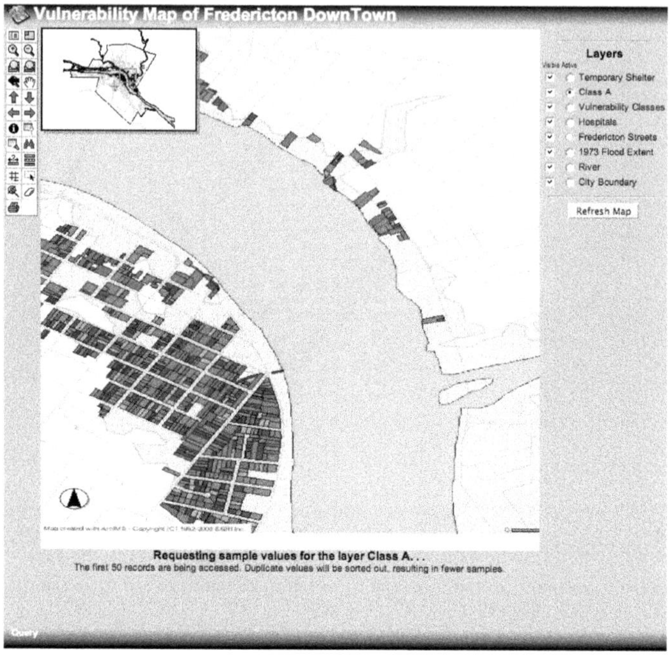

Figure 6: Query example for class A.

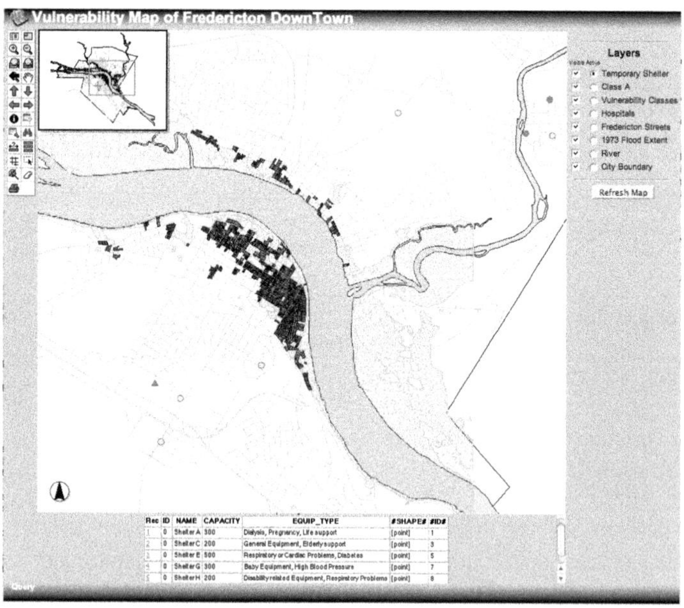

Figure 7: Query for the facilities available in temporary shelters.

6 Conclusions

The results of this research provide a prototype of the web-based application for Fredericton's flood warning system, where accurate and timely warnings can be achieved, and thus maximizing the response time for flood plain residents and emergency managers and to provide a powerful planning tool for flood prone areas. Furthermore, proper evacuation routes can be planned well on time, when water depths on several streets likely to be affected by the floods are known. The idea was to bring the results that would show the inundation at a particular level, represented by a color code, for example having to represent the selections of a particular water level with a particular color for clear distinction.

The developed WebGIS application can assist flood warning practitioners as well as individual people to query the website, for flood evacuation routes; as such the website would allow them to identify evacuation routes as well as to have an idea of the water levels on flooded roads. The flood warnings application is available online and readily accessible to a broad audience to increase the level and quality of services provided.

Using the results of hydrological modelling the system has the capability of identifying the properties that may be affected by an upcoming flood. This can be very helpful to flood practitioners to have an idea of the properties likely to be affected by flooding.

Furthermore, the mapping of socially vulnerable population affected by the flood is presented. Population evacuation in times of disasters is a very important factor and mapping of socially vulnerable population in those areas could go a long way in helping emergency planners in their efforts to evacuate people. This mapping combined with physical hazards and other warning measures have important roles in saving lives and also helping people recover after disasters.

Acknowledgements

This project was financially supported, in part, by the NB Emergency Measures Organization and by the Canadian Department of Natural Resources Geoconnections program as well as by University of New Brunswick and New Brunswick Innovation Foundation (NBIF).

References

[1] Spear MacDonald and Associates Limited, Cumming – Cockburn and Associates, Canada, *New Brunswick Program for a Hydrotechnical study of the Walker Brook Flood plain, Summary Report*, 1983, revised in 1985.

[2] Moreri, K., Mioc, F. Anton, D, B. Nickerson, E. McGillivray, A. Morton, D. Fraser, P. Tang, 2008. *Early warning and mapping for flood disasters, Joint ISCRAM-CHINA and Gi4DM Conference*, Harbin, 2008, 6 pages.

[3] Zipf, A., and Leiner, R., Mobile Internet GIS based Flood Warning and Information Systems, *2nd Symposium on Location Based Services and TeleCartography*, Vienna, Austria, 2004.

[4] Shea, G. Y. K., *A Web-Based Approach to the Integration of Diverse Data Structures for GIS*, M. Eng. Thesis, The University of New South Wales, 2001.

[5] Hansjurgens, B., Megacities, Risk and Social Vulnerability, Helmholtz Association, Retrieved November 13, 2008 from http://www.ehs.unu.edu /file.php?id=334, 2007.

[6] City of Fredericton, Retrieved December 2, 2008 from http://www.fredericton.ca/en/publicsafety/2008April1FloodWatch.asp, 2008.

[7] Canadian Broadcasting Corporation, *CBC News*, Retrieved online on December 2, 2008 from http://www.cbc.ca/canada/new-brunswick/story /2008/04/30/nb-flood.html, 2008.

[8] Nkhwanana Nyaladzani J., D. Mioc, 2009. Early warning system for evacuation of socially vulnerable population during flooding, *Joint Symposium Cartography and Geoinformatics in Early Warning and Crises Management, Towards Better Solutions*. January 19-22, 2009 in Prague, 6 pages.

[9] Watson, C., Using GIS Data for Emergency Preparedness and Response, Retrieved November 30, 2008 from www.geoplace.com, 2008.

[10] Department of the Environment, Water Resources Branch, Fredericton *Planning Region Water Resources Review, Report I – 8001*, 1980.

[11] Andrews, J. (Ed.), *Flooding, Canada Water Book*, Canada Communication Group, 1993.

[12] Butts, M.B. and Khatibi, R., *Workpackage 7: Operational End-user Requirements* – General Functionality deliverable 7.1-Part 1, (FLOODRELIEF project), http://projects.dhi.dk/floodrelief, 2003.

[13] Gibson R. and Erle S., *Google Maps Hacks: Tips and Tools for Geographic Searching and Remixing,* O'Reilly Media Inc., 2006.

[14] Cutter, S.L., Boruff, B.J., Shirley, W.N., Social Vulnerability to Environmental Hazards, *Social Science Quarterly* **84 (2)**, pp.242-261, 2003.

[15] Chakraborty J. *et al.*, Population Evacuation: Assessing Spatial Variability in Geophysical Risk and Social Vulnerability to Natural Hazards, *Natural Hazards Review*, ASCE, ***6*(1)**, pp.23–33, 2005.

Methodological proposal for assessment and adaptation in oil industry plants in Brazil's coastland

G. B. M. Lacerda[1], C. R. G. Souza[2], C. Silva[3] & M. A. V. Freitas[1]
[1]*Program of Energy and Environmental Planning, Coordinated Postgraduate Engineering Program, Federal University of Rio de Janeiro, Brazil*
[2]*Postgraduate Program of the Department of Geography, University of São Paulo, Brazil*
[3]*Program of Civil Engineering, Coordinated Postgraduate Engineering Program, Federal University of Rio de Janeiro, Brazil*

Abstract

This work presents a methodological proposal for medium-term integrated planning and management of extreme natural risks and technological risks, focusing on events of coastal flooding caused by sea level rise (SLR). The research site contemplates private oil industry facilities, located in Redonda Island, Guanabara Bay, Rio de Janeiro, Brazil.

Guanabara Bay suffers interaction with extra-tropical cyclones in the South Atlantic, where disasters, floods and coastal erosion need to be foreseen and managed properly. Under extreme meteorological-oceanographic conditions, a framework of predictive systems for risk management that involves climatic and environmental issues promoting greater capacity and responsiveness are necessary.

In Brazil, the work is justified by the lack of methodological tools ad hoc for assessment and adaptation of coastal structures to face global change vulnerabilities, making them resilient and integrated into the natural coastal systems, and also to the emergency and responsiveness public and corporate systems' plans and programs.

Using software ArcGIS 9.3, simulation scenarios of sea level elevation and flooding impacts were elaborated, whose boundary conditions considered the

WIT Transactions on Ecology and The Environment, Vol 159, © 2012 WIT Press
www.witpress.com, ISSN 1743-3541 (on-line)
doi:10.2495/FRIAR120071

predictions of the IPCC (Intergovernmental Panel on Climate Change), Stern Report, in addition to the syzygy tides and storm surges that occurs in the bay.

The development of the methodology culminated in: the assessment of vulnerabilities of the industrial infrastructures; prior knowledge of flooding from altimetry tracks; the estimation of losses and costs due to physical property damage; the elaboration of thematic maps of adaptive measures and proposed structural interventions to decision making; and the need to prepare a regulatory framework to ensure integration with the plans and programs of public and enterprise systems.

The application of the methodology can ensure the continuity of the company's activities to some level of flooding on the island, or to promote the search for new alternatives of production, storage and logistics.

Keywords: climatic vulnerabilities, assessment, adaptation, methodological proposal, response actions, simulation of scenarios, thematic maps, decision making.

1 Introduction

IPCC (Intergovernmental Panel on Climate Change) [5, 6] have published reports that stipulate sea level elevation up to 1m (meter) till the end of this century. Tide gauge measurements performed in the north coast of the Indian Ocean confirm these predictions, as well as the tendency to rise that has been accelerating since 1870, which currently reaches static sea level elevation values ranging from 1 to 2mm/year. In their studies for up to 100 years, Church *et al.* [2] and Rohling *et al.* [13] point out that sea level rise can reach 1.60m.

Lacerda *et al.* [8] cite that, with the predictable increase in sea level, it is possible to presume the occurrence of great physical and financial losses: infrastructures can become vulnerable and collapse; coastland constructions will disappear; seaports will be destroyed; populations and commercial and industrial installations will have to be relocated; public and private sewage and drainage systems will be affected; new hurricanes could reach the country's south-southeast coast; rainstorms even more intense will punish cities with great social and political impacts over populations in risk areas and; higher and extreme temperatures in short time periods, as well as more diseases, will be registered.

In Brazil, private and public policies on integrated coastal impact management over industrial structures incorporating evaluation elements and structural and non-structural adaptive interventions that have an environmental issue of natural hazards and the institutional integration of technological risks as cross sections to the problem of coastal flooding are still incipient.

For this purpose, an ad hoc instrumental methodology was developed, aiming for an evaluation and adaptation in a diagnostic, prognostic and management context for prevention and reduction of damages arising out of the probability of occurrence of potential flood risks in coastal oil plants. In this work, an integration of information and the rigging of defence and meteorological systems and of plans, programs and emergency systems and actions to existing responses and others to be developed, both corporate and public, are contemplated.

2 Objectives and work pre-viability

The situation or event that overcomes local capacity, causing great hazard, destruction and human suffering and which needs external help is defined by ISDR [3] as disaster. Disasters are classified in 3 groups: hydro-meteorological, geophysical and biological, in which, in accordance to Tucci [16], can cause flooding, draught, impacts on human health, landslides, avalanches and hunger. So forth, natural disasters are earthquakes, floods, tsunamis, etc. Human disasters are those manmade, such as wars. Technological disasters are those derived from accidents involving chemical substances, or dangerous equipment. Considering the actual and future possibility of higher magnitude and frequency on the occurrence extreme natural flood disasters on coastal oil plants, there could happen a conjugation of a natural disaster with a technological, on behalf of installation flooding causing ruptures, leakages, contamination, explosions, harm to the labor force, interruption of supply, environmental fines, among others.

In the point of view of Lacerda and Andrade [7], the management of disasters implies in the development of strategies of actions relative to the governance of extreme natural and technological risks, that consist in the existence of adequate and efficient public policies in the identification of risks through diagnosis, prognosis and management, and in the development of knowhow on risk factor reduction in addition to the preparation for effective response and recuperation.

The main objective of this research consists in the search for answers to the following questions: 1) How to evaluate and know the probable hazards to the vulnerabilities encountered? 2) How to plan to meet the necessary adaptations? 3) Which adaptive ad hoc measures on the island will be taken? 4) Considering that the properties (or assets) of the island must be protected, how to provide its utilities and assure industrial activity till certain level of flooding?

The relevance and innovation of this research relates to the application of a broader methodological proposal than the current studies. An estimated evaluation of the percentage of losses and damage costs, highlighting the need for further studies for preparation of ad hoc thematic maps that point out adaptive measures with a structural intervention proposition was developed. The results of the methodological proposal can be applied in densely populated or strategic coastal sites situated in vulnerable marine flooding regions.

This research presents pre-viability because it considers: a) projections of global occurrence of weather events becoming more intense and frequent, with regional and local consequences; b) the promotion of resilient cities and sites, providing their adaptations to global climate change; c) the need for medium and long term integrated planning and management, to meet the frequent new demands in monitoring and response to flood catastrophe and hazards, with direct relation to projections and events arising from sea level rise; d) early development of adaptive methodological tools predictive to climate change and vulnerabilities resulting from sea level rise in coastal areas for public and / or private managers and decision makers.

3 Characterization of the study area – Redonda Island and Guanabara Bay

The building installations and oil sector infrastructures are located in both coastline and islands in a region of Guanabara Bay in the State of Rio de Janeiro. In this area, the existing emergency and technological risk management programs, as well as medium and long term development plans, do not contemplate environmental risk governance on global climate change, whose effects include local and regional scales.

The research site, Redonda Island, contextualizes the geographic and thematic study by presenting features of flood probability by sea level rise, and by its strategic relevance in regional energy supply. In Rio de Janeiro, situated in Guanabara Bay, the Redonda Island Waterway Terminal (figure 1) and "Ilha d'Água" form together an industrial plant projected for a daily production of 14 million m^3 of different types of oil and gas products.

In figure 1, Redonda's Island existing infrastructure is also presented, as well as a flood simulation inferred from altimeter tracks related to sea level rising from 1.40 m to 19.00 m.

Figure 2 presents an updated panoramic view of the island.

The basic fact that supports the location of the oil industry in Guanabara Bay alleged by TRANSPETRO (marine transport enterprise of Petrobras – Brazilian

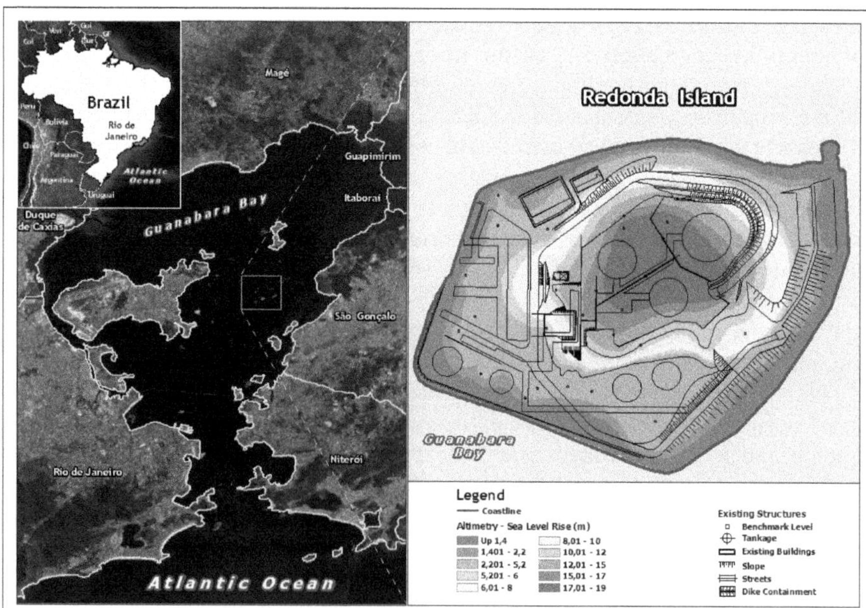

Figure 1: Localization of the study area: Guanabara Bay and Redonda Island, presenting its existing infra-structure and flood simulation by altimeter tracks under rising sea level.

Figure 2: Redonda island terminal.

oil company) is the logistic readiness that the area offers for the transportation of liquefied natural gas stoked in Redonda Island and "Ilhas D'Agua", through a net of gas pipelines set in the Southeast region of Brazil.

Considering the coastal occupation in Rio de Janeiro City and its delimited shoreline, it is necessary to preview critical scenarios and elaborate legal normative instruments to all vulnerable regions – state and municipal, with the establishment of vulnerable climatic zones, adaptive and mitigation measures (structural and non structural) with definition of criteria and terms for occupation and/or evacuation, as well as to ensure the continuity of the island's activities with the least possible socio-environmental, financial and corporative impacts.

In accordance with different sources [1, 10, 14], there are various peculiar characteristics in Guanabara Bay. It is located in the southeastern region of Brazil (Latitude 22°48'12.70" S; Longitude 43° 7'5.62" O), and has an extension of 380km² and is composed of a number of 30 islands and formed by sedimentary plains that incorporates various rivers and has an approximate population of 7.6 million in the surrounding area. The tides have a small and irregular amplitude with a maximum of 1.40m, average syzygy tide of 1.20m, average square of 0.80m and minimum amplitude of 0.20m. The sea level elevation can vary from year to year, with approximate cycles of 20 to 30 years and oscillations of amplitude from 0.10m to 0.50m depending on site and season.

The influence of meteorological tides causes a natural increase in sea level, along with the approach of large waves and storm surge produced by extra tropical cyclones in the South Atlantic, can make the Metropolitan Region of Rio de Janeiro more vulnerable with the occurrence of tragedies and disorders for the population.

Variations associated to climatic events (meteorological tides) have been registered, whose differences between maximum and minimum values can reach 0.90m. Storm surges last in average 5 days in Rio de Janeiro's shoreline. Meteorological tides have been more frequent each time and waves can reach

4.5m. Extreme rainfall events from storm systems associated with the displacement of cold fronts cause precipitations of great intensity and higher frequency, like the ones that occurred in April of 2010 in Rio de Janeiro City, with precipitations of 323mm in a period of 24 hours causing 233 deaths, 3.000 homeless and 11.000 displaced. In January of 2011, in the mountain region of Rio de Janeiro, precipitations reached 222mm in 12 hours, causing 900 deaths, more than 9 thousand homeless and 11 thousand displaced.

The region, embedded between sea and mountains, has a topography that favors flooding and with the effects of a large rainfall, it becomes predictable. In this way, the gravity of the problems varies in function of the storm intensity and of the measures that will be adopted – prevention, hazard reduction and monitoring. The municipal Civil Defense disposes of 2 monitoring systems and of predictive actions for storms that could lead to floods, landslides and destruction: the Alerta Rio (a system that integrates state and local response agencies) and the Operations Center GEO RIO - municipal foundation that manages the slopes of the city. This foundation has installed sirens in some mapped risk areas and uses weather radar capable of detecting thunderstorms 48 hours in advance.

4 Methodological proposal for assessment and adaptation for sea level elevation in coastland oil industry plants

Lieber and Romano Lieber [9] define risk as not being a mere calculation of probability, but a social construction appropriate for the exercise of power, that dictates what is dangerous or not. Risks point out expectations of occurrence of involuntary events, uncertain and hazardous. Its analysis only makes sense if there is a probability of loss. Natural risks are those which are perceived and accepted by a social group or individuals subject to a possible action of a physical progress. Technological risk is defined as the potential occurrence of harmful events to life, as a result of investment decisions in the production structure. Industrial risks, in turn, are intended to clarify two basic components: the estimated frequency of accidental events in the industries and their potential consequences, which latter factor links risk analysis to the emergency response system, standing as an element of impact reduction.

Varnes [17] considers that an adapted concept of analysis and assessment of Risk (R) involves 3 phases: identification of the peril (H = Hazard); assessment of the degree of exposure to danger or degree of incapacity to deal with the consequences of danger (V = Vulnerabilities); and evaluation of response to danger or element at risk (E = Element ay risk), that may depend on the resistance (ability to avoid environmental disturbances) and resilience (ability to respond to the consequences of environmental disturbances), such that R = H. V. E. In this perspective, the climate vulnerability can be defined as the degree of susceptibility of individuals or systems, inability to respond to the adverse effects of climate change, including climate variability and extreme events.

According to the IPCC [6], adaptation in the context of climate change is the adjustment of the natural, social and economic systems in response to the current and future climate stimulus and / or its impacts, in which may be adverse (damage) or beneficial (opportunities). Such a concept of adaptation is an approach that requires the integration and mainstreaming at all levels of decision making in different institutional levels: internal to the institution, between the different sectors of institutions, and between the institution and other sectors of society.

Werner and Smith [18] analyze that in situations of risk, being resilient is not to be invulnerable, which is not to say that in other circumstances the individual would not subdue. Otherwise, it is the capacity to rebuild after being struck, to adapt positively to what has been imposed, drawing experience from difficult situations, enriching in a unique way the experience and then use that learning to reverse the situation in your favor. The term "resilience" has been used as the capacity to face, overcome and be strengthened or transformed by experiences of adversity. This term brings into discussion the issue of governance in an institutional dimension and the social, economic, political and cultural limits of anthropogenic adaptation to ecosystem alterations. For the IPCC [4], a resilient city is the one that developed systems and skills to be able to absorb future shocks and still maintain essentially the same function, structure, systems and identity, getting ready to mitigate consequences of tensions and future shocks. The best way to deal with the problem of climatic vulnerability is to learn to cope and adapt to them (Lacerda *et al.* [8]). Thus, when subjected to shocks and stress, a city or site becomes resilient when they have the ability to adapt and recover, and while maintaining its functions, should also have the ability to make less troublesome future vulnerabilities.

The studies in the development of the methodological proposal *ad hoc* were divided in 3 distinct phases: diagnostic, prognostic and management. Initially, we started with the identification and preliminary assessment of the physical vulnerabilities of the site. For the prior knowledge of flooding by altimetric tracks, scenarios were developed using the ArcGIS 9.3 software, totalizing three different simulation scenarios of sea level rise and its impacts, with boundary conditions pre-established from data prediction of the IPCC (UN), the Stern Review [15], and measurements of syzygy tides and meteorological tides in Guanabara Bay. After that, it was possible to perform an analysis of flood risk

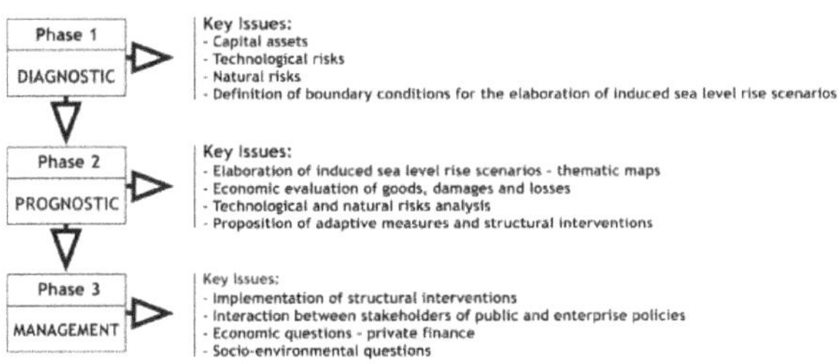

Figure 3: General flowchart - structure and levels of the actions of assessment management and adaptive measures.

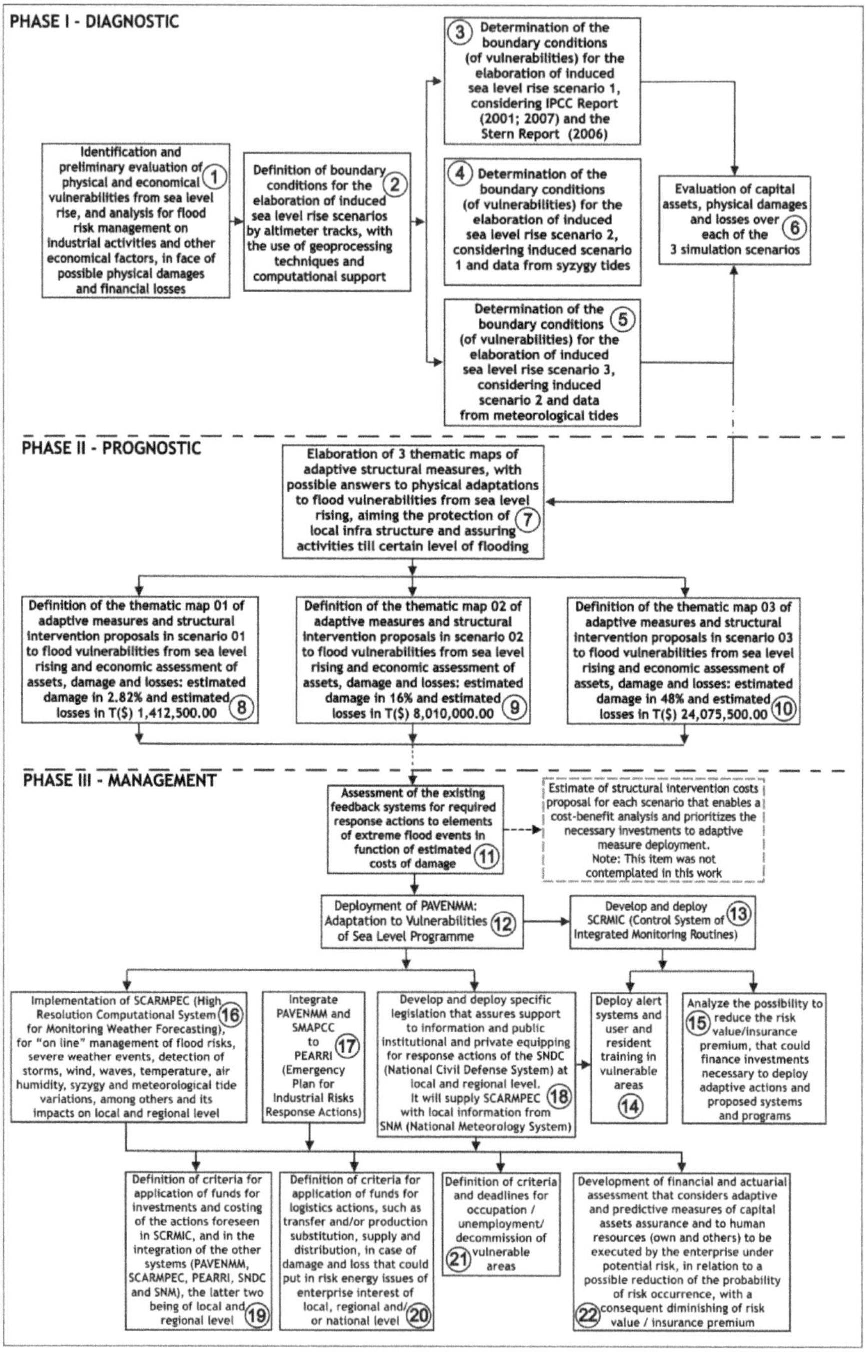

Figure 4: Flowchart of development of the methodology for assessment and adaptation of industrial oil plants.

management on industrial activities, from an economic viewpoint of potential hazard physical loss and property damage. The perception of the need to develop studies to estimate costs for implementation of adaptive measures of structural interventions proposals points to the ongoing research.

Two flowcharts relative to the methodological approach proposal are presented bellow, figures 3 and 4, shown as development board block diagrams.

5 Adaptation and management of extreme natural and technological risks

Besides the evaluation of the technical, meteorological and oceanographic conditions on the research site, additional studies are suggested for structural actions adaptations not only for better understanding of the problem, but for better design solutions. Future studies should cover the estimated costs of the works of adaptive measures and proposed interventions to reduce or avoid losses in the physical facilities, as well as being supported by specific legislation to ensure the integration of all systems, plans and programs cited in the proposed methodology. About the management of extreme natural and technological risks, the PAVENMM has a bias integrator and encapsulates the plans, programs and emergency response systems to be developed and existing. The isolated management of PAVENMM will not ensure the necessary effectiveness and outcomes. As an example of the progress of the work herein suggested, in figure 5 the 3rd scenario is presented, considered the most critical, with the proposition of adaptive structural measures, corresponding to that described in Box 10 of the figure 4.

Table 1 shows, in an arbitrary currency named Tupi ($), estimated value of heritage property, estimated value of the damage, estimated percentage of the physical losses.

Table 1: Estimated values corresponding to good, damage and physical losses.

Scenario	Total Estimated Value of Goods (T$)	Total Estimated Value of Damage (T$)	Estimated Percentage of Physical Losses
III	50.105.000,00	24.075.500,00	48,00%

To reduce or prevent losses in the physical facilities of the plant, this paper proposes the implementation of structural measures proposed in drawings in type of croquis. To date it has not been possible to estimate in the present work, the cost of works and proposed interventions to evaluate the cost-effectiveness.

5.1 Scenario III

In this third scenario, the occurrence of an extreme natural event was considered beyond the events of sea water elevation, estimated in IPCC and Stern reports, and the syzygy tide. This extreme event was a 3m high wave which entered

Guanabara Bay in April of 2008, and was considered a rare event. However, similar marine disturbances occurred in October of 2008, confirming the higher frequency of this extreme event. In this case, sea water elevation can go up more than 5m, making today's use of the island unfeasible.

This scenario may need the adoption of more complex measures which will demand more specific studies by the company that owns the island since estimated loss percentage is 48.00% and the total amount of damage is valued in R$24,075,500.00, and also because the only part of the island above sea level will be from elevation quota 9.00 to 19.00. The implementations of some measures to reduce or avoid losses in the physical facilities of the plant are suggested for scenario III, as follows (see Figure 5):

1. Protection and lifting of slopes of the eastern part of the island;
2. Protection of the other sectors of the island with conventional concrete walls (profile L) 5.50 m high;
3. Creation of 2 drainage wells in the island for pumping rain water (if the drainage system of the island is flooded);
4. Forecast of a system of drainage sluices to prevent flooding of the island;
5. Lifting of public area (i.e., streets, sidewalks) in the south and west of the island (for concurrent event of heavy rain);
6. Forecast deployment of FPSO (Floating Production, Storage and Offloading) structures sheltered in breakwaters, for non interruption of existing processes on the island, such as loading and unloading of gas;
7. Creation of breakwater structure in "L" to protect the island and give shelter for FPSO vessels.

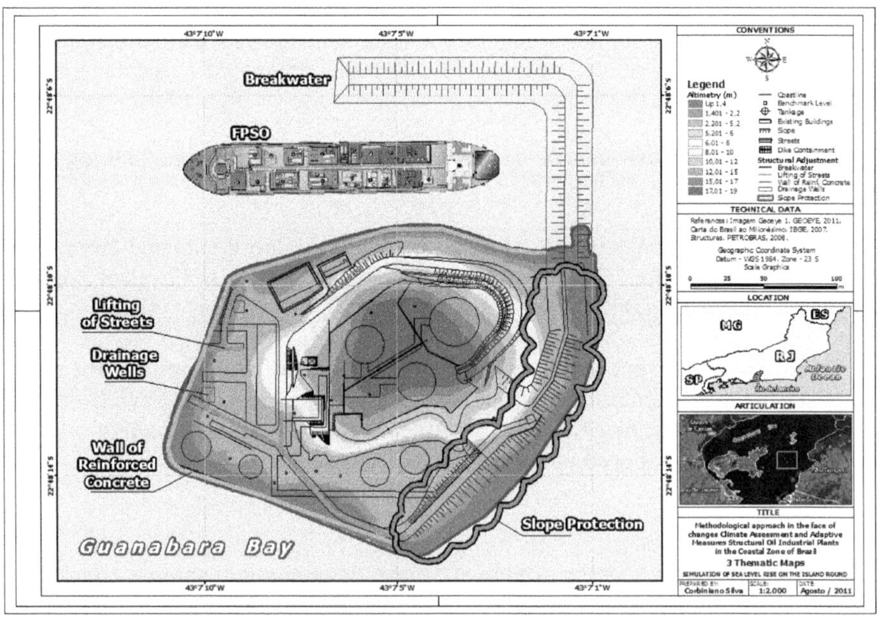

Figure 5: Proposed interventions in scenario III.

Scenario III: Simulation of sea level elevation in Redonda Island, considering syzygy tide, analyzed reports and the occurrence of extreme natural events.

6 Conclusions and recommendations

Ripley [12] emphasizes that with the increase in catastrophic evidences and events, the importance of transforming common folk in better survivors stands out. Having in mind that you should not trust being lucky, and that in the middle of a crisis it is too late to learn lessons: part of the resources designated to security, plans, programs, politics for calamities and reactions must also contemplate participation and training of social agents and the population at risk.

The lack of clear and unequivocal knowledge systems about environmental problems as major issues underlying the growing social perceptions of risks makes them more socially problematic and intolerable. Coastal real estate prices will probably be one of the first economic indicators to reflect the rise in sea level, where investments in beach properties will be most affected.

In the case study of Round Island, it is assumed that the proposed methodology should address, in a more restrictive form, an analysis under the criteria of cost and benefit. In case the enterprise owner decides to protect their assets and extend their usefulness to the point of securing the industrial activities at the site till a certain level of flooding, it should have resources to spare to implement measures necessary for physical adaptations to the island, obtained in future studies, which could effectively mean introducing the environmental variable in the integral management of industrial risks in the plant. In order to reduce climate vulnerability and consequent risks to populations and activities, and integrate these criteria into PAVNMM, a more effective local legislation must consider the change in zoning for the purpose of land use planning and the types of activities allowed in densely populated urban areas, also covering other areas of expansion, metropolitan and micro-regions through Master Plans. In Brazil, the approach on flood risk management, despite being the subject of several important studies, is not contemplated in the National Policy for Water Resources - Law 9433/97 [11], and disregards a regulation in order to position decision makers at all stages: (1) pre-event actions of prevention and protection, (2) actions during the event, for protection and aid, (3) actions post-event of area recovery activity and rehabilitation, (4) mitigation, parallel to the others, which consists of risk management and loss reduction policies. This legislation must provide for and regulate the decommissioning of industries or other activities under potential risk in vulnerable regions which could cause socio-environmental damage, hazards on logistic supply and tourism, among others.

Omitting preventive actions without studying, evaluating and predicting adaptive ad hoc measures, establishing their risks and their mitigation solutions and responses to emergencies can constitute in loss of material and of human lives. Economic losses arising from recovery of structures, environmental fines, recovery of degraded areas, replacement of supply by business interruption, claims for damage to property, to human lives and to the image of the company owner may also be significant. The reduction in the likelihood of risk and the consequent reduction of insurance premium would also be affected. Insurance companies may suggest that the adjustment measures to vulnerabilities be implemented in areas under potential risk of flooding.

References

[1] Amador, E. S., 1997. Baía de Guanabara e Ecossistemas Periféricos: Homem e Natureza. Rio de Janeiro: E. S. Amador, 1997.

[2] Church, J., Gregory, J., Huybrechts, P., Kuhn, M., Lambeck, K., Nhuan, M.T., Qin, D., Woodworth, P., 2001. Understanding 20th century sea-level rise and projections for the future. Alfred-Wegener-Institut für Polar- und Meeresforschung e PIC - Electronic Publication Information Center.

[3] ISDR (International Estrategy for Disaster Reduction). Terminology: basic terms of disaster risk reduction. (revisão de 31/03/2004). Disponível em: http://unisdr.org/eng/library/lib-terminology-eng-p.htm

[4] IPCC, 1994. Working Group II Report - "Impacts, Adaptations and Mitigation of Climate Change: Scientific-Technical Analyses", Contribution of Working Group II to the Second Assessment Report of the Intergovernmental Panel on Climate Change.

[5] IPCC, 2001. Climate Change, The Scientific Basis, Cambridge Univ. Press, pp.881

[6] IPPC, 2007. Climate Change 2007: The Physical Science Basis - A report of Working Group I of the Intergovernmental Panel on Climate Change – Summary for Policymakers. p.18, Cambridge University Press, Cambridge, United Kingdom and New York, NY, USA. IPCC, 2007. AR 4, Climate Change 2007: Impacts, Adaptation and Vulnerability, Contribution of Working Group II to the Fourth Assessment Report of the Intergovernmental Panel on Climate Change.

[7] Lacerda, G., Andrade, E. 2009. Mudanças climáticas e Análise de Risco da Indústria de Petróleo no Litoral Brasileiro. Fundación Mapfre, Madrid.

[8] Lacerda, G., Mamede, E., Assad, L., Freitas, M., 2010. Hydrological Vulnerabilities from Sea Level Instability: Physical Action Adaptation and Risk Management. Storm Surges Congress 2010. University of Hamburg, Germany.

[9] Lieber, R. R.; Romano Lieber, N. S.. O conceito de risco: Janus reinventado. In:Minayo, M.C.S.; Miranda, A.C. Saúde e ambiente sustentável: estreitando os nós. Rio de Janeiro: Ed. Fiocruz/Abrasco, 2002. p.69-111.

[10] Neves, C., Muehe, D., 2008. Impactos Possíveis da Elevação do Nível do Mar e Eventos Climáticos - RIO PROXIMOS 100 ANOS – O aquecimento global e a cidade. Prefeitura da cidade do Rio de Janeiro – Secretária Municipal de Urbanismo – Instituto Municipal de Urbanismo Pereira Passos, 2008.

[11] POLÍTÍCA NACIONAL DE RECURSOS HÍDRICOS, Lei 9433/1997.

[12] Ripley, A., 2008. Impensável: Como e por que as pessoas sobrevivem a desastres. São Paulo: Globo.

[13] Rohling, E.J.; Grant, K.; Hemleben, C.H.; Siddall, M.; Hoogakker, B. A. A.; Bolshaw, M. and Kucera, M., 2007. High rates of sea-level rise during the last interglacial period. Nature Geoscience (published online in 16/12/2007, doi:10.1038/ngeo.2007.28).
[14] Santos, C., Silva, Maria., Salvador, Marcus., 2004. Dinâmica Sazonal e os Efeitos das Ressacas nas Praias de Niterói/RJ. Revista Brasileira de Geociências.
[15] Stern, N., 2006. Stern Review. The Economics of Climate Change. UK, p.267.
[16] Tucci, Carlos., 2007. Inundações Urbanas. – Porto Alegre: ABRH/RHAMA.
[17] Varnes, D. J., 1984. Landslide Hazard Zonation: Review of Principles and Practice. Paris: UNESCO Press, p.56.
[18] Werner, E., Smith, R., 1982. Vulnerable but invincible: a longitudinal study of resilient children and youth. New York: MacGraw-Hill.

Section 3
Emergency preparedness and response

Floods emergency management: the value of potential and actual damage estimation

D. Molinari[1], F. Ballio[1] & S. Menoni[2]
[1]*Dept. of Environmental, Hydraulic, Infrastructures and Surveying Engineering, Politecnico di Milano, Italy*
[2]*Dept. of Architecture and Planning, Politecnico di Milano, Italy*

Abstract

Emergency management (intended here from emergency planning to reaction) is a crucial aspect of risk management, allowing us to handle residual risk. Therefore, its economic convenience is an important piece of information in designing flood risk management strategies. This paper seeks to answer the following questions: How expected damage reduction due to emergency management can be estimated? Which analyses are required? Are there available tools in literature?

In the paper, a framework is provided to describe the process of emergency management and preventive actions leading to damage reduction. The latter are differently handled according to whether they are addressed to reduce the hazard or exposure/vulnerability. The framework is then applied to a real case study. Results suggest that damage reduction estimation is possible, even though available models still require further improvement, in terms of transferability and required data/knowledge. On the other hand, results show the utility of the information supplied by this kind of analysis for decision makers.
Keywords: emergency management, EWS, depth-damage curves, actual damage, floods directive.

1 Introduction

Looking at the past, history teaches us that where a risk is present, it is impossible to totally protect communities from the likelihood of a disastrous event even if proper mitigation measures are implemented; as a consequence, a "residual risk" always exists which affects exposed systems. Emergency

doi:10.2495/FRIAR120081

management (intended here from emergency planning to reaction) allows us to handle residual risk; impacts can be decreased if people are aware of upcoming events and are prepared to react. For this reason, the role of emergency management became increasingly important in the last few decades [1, 2]

Although the prime purpose of emergency management is to save life, which ideally always justifies the adoption of emergency management strategies, this paper focuses upon the economic savings caused by emergency management, with particular attention on floods. Economic convenience is a critical piece of information in defining flood management policies; the recent Floods Directive [3], for example, explicitly requires us to take account of costs and benefits in the design of flood risk management plans.

When damage to people is not considered, some questions arise: How expected damage reduction can be estimated? Which analyses are required? Are there available tools in literature? This paper seeks to answer these questions.

A methodology is supplied and damage reduction is actually estimated for a case study. Despite all the uncertainties that presently affect case study's results, the implementation of the methodology allows us to take a further step forward in the comprehension of the problem at stake; on the one hand, it allows us to identify which are the current difficulties in applying the methodology and, as a consequence, which are the priorities for future research. On the other hand but even more important, it allows us to understand what the value of damage reduction estimation is in terms of knowledge supplied to emergency managers. Indeed, a principle is maintained in the paper according to which good emergency management should assemble feasible and effective strategies to cope with the event which are specific for the context under investigation. The evaluation of expected damage reduction is thus a critical piece of information to adjust/design emergency plans, by defining better response strategies.

The paper is organized as follows: In section 2 a framework is provided to describe the process and preventive actions leading to damage reduction. The framework is linked to available modeling approaches for the estimate of expected damage reduction. The case study is presented in section 3 which explains how damage reduction can be estimated in practice; results are then discussed, with particular attention on the utility supplied by the assessment in terms of available knowledge for emergency managers. A critical analysis of the methodology is supplied in section 4. Finally, section 5 reformulates findings from previous sections in order to answer the original questions of the paper.

2 Emergency management in case of flood: The aim of reducing potential damage

Figure 1 depicts the emergency management process as conceived in this paper:

- When a flood is monitored or forecasted, civil protection makes choices, ideally according to what has been previously arranged in the emergency plan; these choices regard both warning and preventive measures which aim at reducing exposure, vulnerability and flood intensity.

- Once the warning has been issued, lay people react to it and implement, in turn, preventive measures. The latter are addressed to reduce exposure and/or vulnerability; literature highlights that people reaction depends on both warning features and social vulnerability [2, 4, 5];
- The extent of items at risk (after preventive measures have been implemented), their vulnerability (from all its perspectives: social, institutional, functional and physical) and flood intensity concur then to define expected damages.

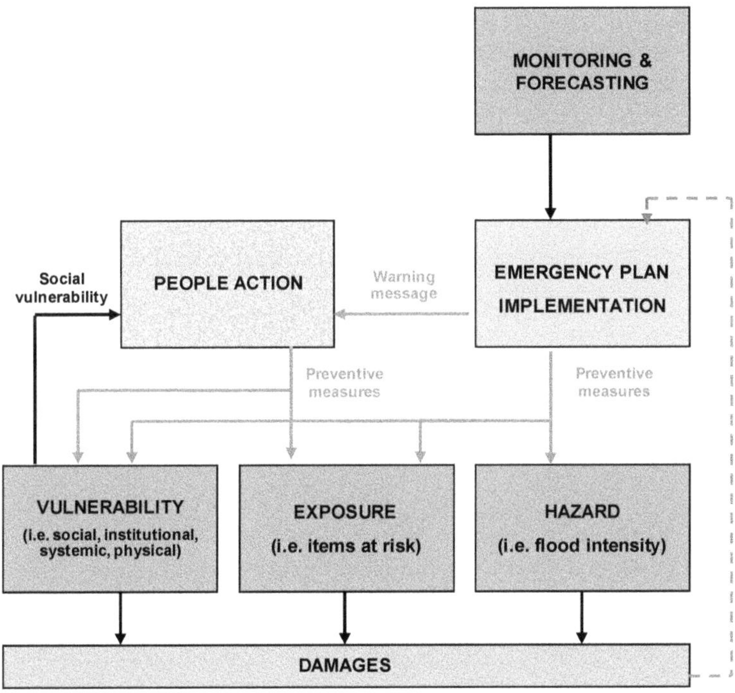

Figure 1: The emergency process, as conceived in this paper. Coloured arrows represent preventive measures implemented during emergency, whose effects need to be modelled.

Evaluating damage reduction due to emergency management is equivalent to modelling the effect (on expected damages) of preventive measures implemented during the event, by both civil protection and lay people. Such an assessment can be used then to adjust emergency plans; this is pointed out in fig. 1 by the back-going dotted arrow which stresses how assessment's results can influence initial choices.

Referring to the most quoted terminology in available literature [6–8] the problem of assessing damage reduction can be viewed in terms of the estimate of actual versus potential damages where potential damages are those damages that might happen when no emergency management exists; actual damages are, instead, damages that would really happen after preventive measures are adopted.

Improving potential damages models is beyond the scope of this paper. A variety of tools are already available in literature [9–12]. In a nutshell, potential damages are usually estimated by means of depth-damage curves (which supply expected damage to affected items against water depth), field surveys or back analyses of damage data. Focusing instead on preventive measures, fewer tools are available [13–16]. In this paper, the effect of preventive measures on potential damages is differently modeled according to the objective of the actions; indeed two kinds of preventive actions can be put in place (fig. 2):

- those reducing the intensity of the hazard (i.e. levees rising and reinforcement, bridge gates, temporary dikes, etc.);
- those aiming at limiting exposure and vulnerability (e.g. moving contents, temporary water gates, etc.).

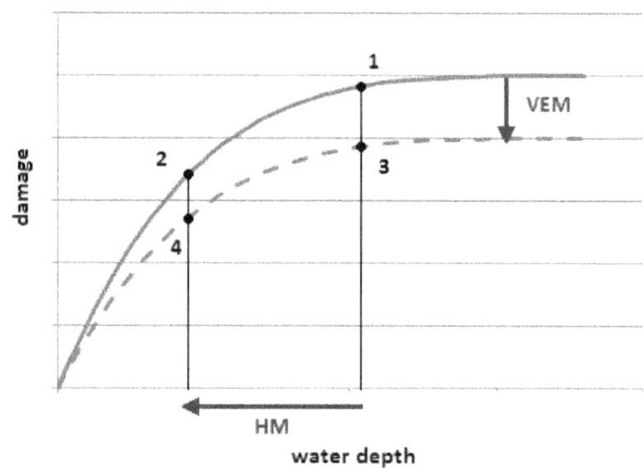

Figure 2: Effect of hazard mitigation (HM) and vulnerability and exposure mitigation (VEM) on potential damages.

In both cases, effectiveness depends on the level of preparedness as well as the available time for implementation; still, the effect on the potential damage curve is different. In the first case, preventive actions imply that, for a given event, the water depth in the flooded area decreases; accordingly, a shift occurs on the x-axis (identified by HM in fig. 2) so that if 1 represents the starting point for potential damages, the new value for damage (i.e. actual damage), after HM preventive actions, is 2. Otherwise, it is possible to explain the effect of actions on exposure and vulnerability as another depth-damage curve describing actual damages. The latter can be interpreted as the consequence of a shift (identified by VEM in fig. 2) in the potential damage curve so that, for a certain water depth, damage is less because of exposure and/or vulnerability reduction. Accordingly, if 1 represents the starting point for potential damages, the new value for damage (i.e. actual damage), after VEM preventive actions, is 3. Clearly, the combination of the two kinds of action brings the original damage to point 4.

Actual damage estimation requires the evaluation of both the effects. As a matter of fact, whilst the effect of preventive actions on hazard reduction can always be estimated by means of a hydraulic analysis, the evaluation of the consequence of preventive actions on exposure and vulnerability requires specific models and is still an open question [2, 17]. In section 3, a case study is supplied to describe how such an assessment can be carried out with available models. Both in the assessment of potential and actual damages, all types of damages (i.e. direct/indirect, tangible/intangible) and exposed items should be ideally accounted for. Nevertheless, damage to people is not considered in this paper, the objective being the evaluation of the economic benefit of emergency management.

3 Case study

The scope of this section is to describe how actual damages can be estimated in practice. Potential damages are first estimated; then the effect of preventive actions on damage reduction is assessed.

The case under investigation is the town of Sondrio, in the Italian Alpine region, where flood risk is induced by the River Mallero. It is important to clarify that the case study does not aim at supplying an exact estimate of potential and actual damages; this would be challenging in the Italian context where specific tools and data for their estimation are not available. On the contrary, the objective is to highlight which tools are available for damage estimation, which kind of knowledge such an estimate could supply (in a plausible case) and how this knowledge could be used to improve emergency management. For this reason, available tools from other countries are here implemented, being aware that their fitness with Italian context must still be proved and represents a priority for future research; as a result of the exercise, the following results are however considered reliable.

Emergency management in Sondrio is responsibility of local authorities who arranged for an emergency plan to deal with flood events. The plan sets preventive actions to be implemented by both civil protection and lay people after a warning is issued. The warning system is based on river discharge thresholds; once flood forecasts exceed one of them, the corresponding warning level is implemented. The plan schedules two main preventive measures in case of warning:

- the use of temporary gates (from now on called “bridge gates”) to close river bank openings at bridges, which is done by civil protection;
- people-individuals’ actions like moving contents, turning-off gas, sandbagging, etc.

3.1 Potential damages assessment

Potential damages estimation in Sondrio focused on the following types of damage:

- physical/direct damage to buildings;

- physical/direct damage to infrastructures;
- emergency costs, defined here as the cost required us to face the emergency.

Indirect damages due to business/services disruption as well as intangible damages were not considered in the present study.

According to section 2, physical/direct damage to buildings and infrastructures was calculated by means of available depth-damage curves; specifically, curves by the US Army Corps of Engineers [19] were implemented for buildings while the Dutch Standard Method [20] was adopted for infrastructures. The two methods were selected taking into account their transferability to the context under investigation that is the correspondence between hazard and vulnerability parameters considered in the methods, the vulnerability of exposed items and the available knowledge about the hazard. A literature review highlighted that few tools are instead available for emergency costs estimation; here the model suggested by Penning-Rowsell *et al.* [6] was implemented which estimates emergency costs as a fixed percentage of the value of physical/direct damage to buildings equal to 10%. The model was validated with data coming from the local database RaSDa (acronym of "Sistema per la Raccolta delle Schede Danni", Italian translation for "System to collect damage data") [18]. Fig. 3-serie 1 displays assessment results in terms of the extent of potential damages against river discharge.

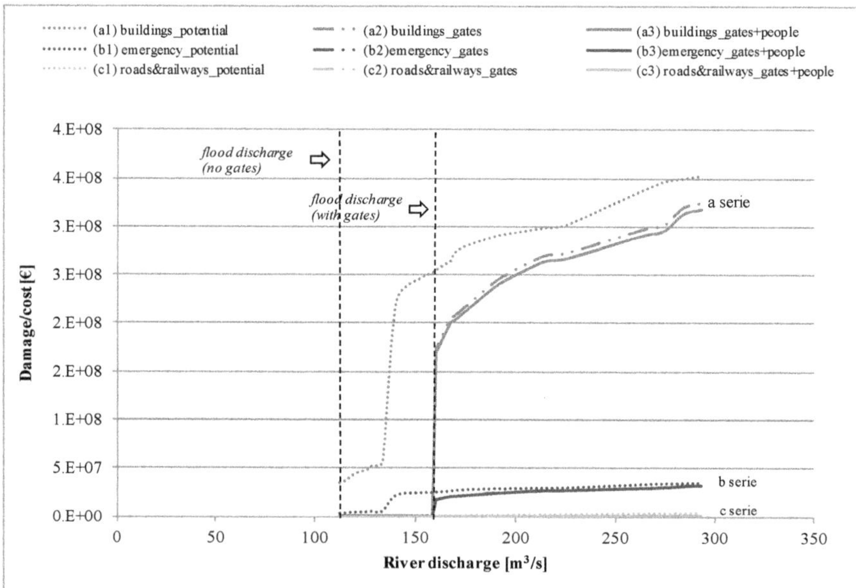

Figure 3: Results of damage assessment: potential vs. actual damages; figure distinguishes between different classes (referred to as a, b, c) and between the effect of bridge gates only and the joint effect of bridge gates and people actions (referred to as 2, 3).

3.2 Actual damages assessment

According to section 2, actual damages estimation requires to assess the effect of the preventive measures scheduled in the emergency plan on potential damages. As regards to bridge gates, an hydraulic analysis was performed (not reported here [18]) given that the effect of bridge gates is to reduce the hazard by increasing the bankfull discharge of the river (see section 2). According to the analysis, the latter increases up to 160 m3/s.

People actions was modelled instead as a shift in potential depth-damage curves (see section 2). In detail, a fixed percentage of potential damages was considered for every water depth value. This percentage is equal to 4.5%, according to the value suggested by [6] for the warning context observed in Sondrio. Fig. 3 reports the comparison between results for potential and actual damages estimation. To be noted that the figure distinguishes between the effect of bridge gates only and the joint effect of bridge gates and people actions.

3.3 Utility of damage assessment results

Analyses carried out in the previous section allow to infer useful information regarding both the system under investigation, its (economic) convenience and its functioning during emergency. Useful knowledge derives initially from potential damages assessment; in detail, fig. 3 picks out that:

- first, physical/direct damage to buildings is significant and represents the major component of the total damage;
- second, physical/direct damage to single infrastructures is less relevant than physical/direct damage to buildings; specifically, damage to infrastructures is, on average, four orders of magnitude less than that to buildings.

Regarding the former point, it must be observed that this evidence does not depend on "emergency management" but it is linked to potential damages; in other words, to reduce damage to buildings, priority must be given to mitigation actions "in time of peace" (i.e. before the event strikes) which would aim at reducing buildings vulnerability (e.g. buildings codes), their exposure (e.g. planning) as well as the hazard (e.g. levees, detention basins). With respect to the second point, instead, the result is sensible, considering that damages to infrastructures are usually mainly indirect (e.g. damage due to the disruption of traffic, economic activities, etc., see section 4).

Actual damages assessment allows instead to infer useful information about emergency management economic convenience as well as on emergency management planning. Fig. 3 puts into evidence that damage reduction due to preventive actions is mainly due to bridge gates. Such an evidence suggests the more suitable emergency strategy too. Indeed, given that benefit from bridge gates is significant and actually with no costs (but for indirect damages due to traffic/business disruption, not computed here, see section 4) emergency managers should act "for safety" in their adoption. In other word, they should always adopt bridge gates when a flood cue is observed (forecasted), even if uncertainty is high. Moreover, it should be borne in mind that, even if damage to people is not evaluated here, the main objective of emergency management should always be safe life. In accordance with this, limiting people access to flooded areas sounds sensible as road users represent up to half of the deaths during floods [21, 22]; from this perspective, the precautionary use of bridge gates is wise. Otherwise, fig. 3 highlights that damage reduction due to people action is limited. Thus, considering also that major costs can occur if false warnings are frequent (also in terms of loss of trust, wolf-crying syndrome, etc.), emergency managers may decide to act "against safety" in warning people, waiting until more certain forecast is available.

4 Critical discussion

Besides inherent uncertainties of damage models [23], the estimation of direct damages suffer from two main limitations. The first relates to infrastructures; present damage estimation does not take into account damage to accessories and installations (e.g. bridges, tunnels, water tanks, purification plants, control room, etc.) which, instead, can be relevant [18]. However, authors are not aware of parametric models which are able to adequately estimate damage to accessories and installation.

A second limitation relates instead to emergency costs; even though costs have been set equal to 10% of damage to buildings, validation with RaSDa data (not reported here [18]) suggests that they could be much more significant. This is especially important when emergency costs are compared with emergency benefits. Fig. 3 shows, for example, that emergency costs (i.e. the "b" serie) and damage reduction to buildings (i.e. the difference a1-a3), which represents the main benefit, are of the same order in the case under investigation. However, because of the above modelling uncertainty, it is possible that emergency costs are higher than benefits, bringing to emergency management ineffectiveness. Thus, improved models to assess emergency costs are required.

When actual damages estimation is considered, the main problem regards instead damage reduction due to people actions which was modelled as a fixed shift in potential depth-damage curves; this is the simplest method but it is worthy to note that, when data are available, assessment methods based on the modelling of people behaviour would represent a more suitable choice [16, 24].

Finally, as regards neglected types of damages, it is sensible that the inclusion of indirect and intangible damages severely affect emergency management convenience, as derived from the case study. From an economic point of view, it is evident that potential and actual damages change when indirect damages are considered. For example, indirect damages due to bridge gates can be relevant given that the economic activity of the area can suffer from traffic disruption. On the contrary, also indirect benefits can be significant; in fact bridge gates allow us to avoid the inundation of the city for $Q < 160$ m^3/s preventing not only direct damages but also indirect damages to enterprises (i.e. loss of income).

On the other hand, it is clear that emergency management convenience goes beyond economic considerations. When intangibles are considered, convenience increases given that emergency management enables us to save lives, memorabilia, cultural heritage, etc. [16, 17]. However, available tools for indirect and intangible damage estimation seem unsuitable for practical and systematical purposes and require specific knowledge and usually unavailable data.

5 Conclusion

This paper analyzes and implements available tools for damage assessment in a pragmatic framework to describe and assess damage reduction due to emergency management; the framework is then applied to a real case study.

The conceptualization of the problem at stake represents the main strength of the paper; section 2 supplies a pragmatic framework to describe the emergency management process and preventive measures leading to damage reduction and suggests how damage reduction modeling should be approached as the objective of preventive actions changes (i.e. reducing hazard vs. reducing exposure/vulnerability).

The case study proves that damage reduction estimate is possible even though present models still require further improvement and must be adapted to the specific context under investigation; this strongly influence present accuracy. At the same time, the case study points out that assessment results supply useful information, not only in terms of economic convenience of emergency management but also in terms of the comprehension of the system under investigation and its functioning during emergency. From this perspective, this paper supports and encourages present efforts of the research community in improving damage models.

References

[1] Romang, H., Zappa, M., Hilker, N., Gerber, M., Dufour, F., Frede, V., Berod, D., Oplatka, M., Hegg, C. & Rhyner, J. IFKIS-Hydro: an early warning and information system for floods and debris flows. *Nat Hazards* **56**, pp. 509-527, 2010.

[2] Modelling the damage reducing effects of flood warnings; FLOODsite project, Report T10-07-12. Online. www.floodsite.net

[3] Directive 2007/60/EU of the European Parliament and of the Council of 23 October 2007 on the assessment and management of flood risks, Official Journal of the European Union, 2007.

[4] Handmer, J.W. & Ord, K.D. Flood warning and response. *Flood warning in Australia*, ed. Smith, D.I. & Handmer, J.W, Centre for Resource and Environmental Studies: Canberra, pp. 235-257, 1986.

[5] Sorensen, J. & Mileti, D., Warning and evacuation: answering some basic questions. *Ind Crisis Q* **2(3-4)**, pp. 195-210, 1989.

[6] Penning-Rowsell, E.C., Johnson, C., Tunstall, S., Tapsell, S., Morris, J., Chatterton, J., Coker, A.& Green, C., *The benefits of flood and coastal risk management: a manual of assessment techniques*, Middlesex University Press: London, 2005.

[7] Disaster loss assessment – Guidelines, State of Queensland and Commonwealth of Australia, 2002.

[8] Framework for estimating national economic development benefits and other beneficial effects of flood warning and preparedness systems, IWR Report 94-R-3. Online. www.iwr.usace.army.mil/docs/iwrreports/94-R-3.pdf

[9] Evaluating flood damages: guidance and recommendations on principles and methods; FLOODsite project, Report T09-06-01. Online. www.floodsite.net

[10] Smith, D.I. Flood damage estimation-a review of urban stage damage curves and loss function. *Water SA*, **20(3),** pp. 231–238, 1994.

[11] Thieken, A.H., Muller, M., Kreibich, H. & Merz, B.: Flood damage and influencing factors: new insights from the august 2002 flood in Germany. *Water Resour. Res.*, **41**, doi: 10.1029/2005WR00417, 2005.

[12] Thieken, A.H., Ackermann, V., Elmer, F., Kreibich, H., Kuhlmann, B., Kunert, U., Maiwal, H., Merz, B., Muller, M., Piroth, K., Schwarz, J., Schwarze, R., Seifert, I. & Seifert, J. Methods for the evaluation of direct and indirect flood losses. *Proc. of the 4th Int. Symp. On flood defense*, Toronto, Canada, 2008.

[13] Day, H. J.: *Flood warning benefit evaluation-Susquehanna River Basin (urban residences), ESSA Technical Memorandum* WBTM, Hydro-10, National Weather Service: Silver Spring, 1970.

[14] Chatterton J. B. & Farrell, S. *Nottingham flood warning scheme: Benefit assessment*, Report prepared for Severn-Trent Water Authority, Middlesex, U.K., 1977.

[15] Handmer, J. W. & Smith, D.I.: Adjustment factors for flood damage curves. *Journal of Water Resources Planning and Management*, **116 (6)**, pp. 843-846, 1990.

[16] Parker, D., Tapsell, S. & McCarthy, S.: Enhancing the human benefits of flood warnings. *Nat. Hazards* **43**, pp. 397-414, 2007.

[17] Carsell, K.M., Pingel, N.D. & Ford, D.T.: Quantifying the benefit of a flood warning system. *Natural Hazards Review*, **5(3),** pp. 131-140, 2004.

[18] Molinari, D.: *Flood early warning systems performance: an approach at the warning chain perspective,* Politecnico di Milano, PhD Dissertation: Milan, Italy, 2011.

[19] Generic depth-damage relationships for residential structures with basements, economic guidance memorandum, US Army Corps of Engineering, Economic Guidance Memorandum (EGM) 04-01. Online. planning.usace.army.mil/toolbox/library/EGMs/egm04-01.pdf

[20] Kok, M., Huizinga, H.J., Vrouwenvelder, A.C.W.M. & Barendregt, A.: *Standard Method 2004: damage and casualties caused by flooding*. Road and Hydraulic Engineering Institute: Netherlands, 2005.

[21] Gruntfest, E. & Handmer, J.W. (eds.). *Coping with Flash Floods. Vol. 77 NATO Advanced Study Institute*, Kluwer Academic Publishers: Dordrecht, 2001.

[22] Jonkman, S. & Kelman, I.: An analyses of causes and circumstances of flood disaster deaths. *Disasters*, **29 (1)**, pp. 75-97, 2005.
[23] Merz, B., Kreibich, H., Thieken, A. & Schmidtke, R.: Estimation uncertainty of direct monetary flood damage to buildings. *Nat. Hazards Earth Syst. Sci.*, **4(1),** pp. 153-163, 2004.
[24] Molinari, D. & Handmer, J.W.: A behavioral model for quantifying flood warning effectiveness. *J. Flood Risk Manag.*, **4(1)**, pp.23-32, 2011.

Flood disaster risk in Jakarta, Indonesia

A. Neolaka
Department of Civil Engineering, Faculty of Technology,
State University of Jakarta, Indonesia

Abstract

This research aims at investigating the level of environmental awareness knowledge on the part of Jakarta community members in anticipating the huge flood that occurs annually in Jakarta city and its neighbouring areas. Survey and field observation have been employed in this research that focuses on the eastern part of Jakarta. A total number of 160 respondents from 40 families have been involved in the study. They consist of parents and their ≥16-year-old sons and daughters who were asked to complete a questionnaire for this research. Findings show (1) a knowledge sub-indicator level of 21.88% for environmental awareness and 43.12% for flood risks; whereas, (2) sub-indicator related to attitude towards environment of the community members living in flood-prone area is 47.25%. This later figure is an average of several indicators. When each of these indicators is observed, we can see a very low indicator of environmental related behaviour; i.e. people who are not disappointed with the existence of illegal buildings along the river flow reach 28.75%. By careful observation of the indicators, we can identify one indicator that signifies very poor environmental behaviour; i.e. the community are not disappointed to see the rows of illegal constructions along the river banks. Interviews with the community, represented by parents, reveal that they understand about the risks of yearly floods, know that they must dispose waste/trash in proper place, and comprehend that they should keep the drainage system in front of their houses free of trash/garbage. However, their knowledge is not reflected in their daily living practices. The Jakarta Community does not have sufficient environmental awareness that their effort to control/manage the risk of floods is very poor. It gets worse by their negative attitude about moving out from the flood-prone areas (43.7%). Findings of this research are supported by what actually happens in Jakarta as flood occurs every time it rains. In conclusion, public environmental awareness in Jakarta, focused on anticipating the occurrence of floods, is quite poor. This has been the

WIT Transactions on Ecology and The Environment, Vol 159, © 2012 WIT Press
www.witpress.com, ISSN 1743-3541 (on-line)
doi:10.2495/FRIAR120091

influence of socio-cultural diversity, difficulty in obtaining employment, and poverty. People do not even have the chance to think about caring for their environment despite the fact that their living practices can bring about disaster to them.
Keywords: flood risk, behavioural, environmental awareness, community of Jakarta.

1 Introduction

Jakarta as the capital of the Republic of Indonesia attracts the world's attention for there is always a flood every time it rains. As the nation's capital, as well as the centre of economic activities, trade, education, industry, tourism and others, indeed, Jakarta has become a major destination for many individuals of varied cultural backgrounds to earn a living. Their different sociocultural backgrounds have impact on the level of concern they have for their environment. This is reflected in their daily life behaviour that does not proactively maintain their immediate living environment. According to sociology theory, a society that consists of social groups of different cultural background, ethnicity, race, ideology, traditions, behaviours, etc., allows the occurrence of disagreements that lead to conflicts and encourage changes in society. Natural disasters such as floods can lead people who live in the affected areas to migrate to a safer place. Sometimes the floods caused by the actions of citizens themselves, for example; cutting down the forest/tree in an area set aside for water catchment, disposing trash is not in place and so on [1]. This sociological theory reinforces the notion that Jakarta with its variety of social, culture, and traditions will strive to keep up with the difficult economic life, frequent conflicts and frictions in life, and lead people to an unintentional lack of concern for their living environment. Those who came to Jakarta to find work and settle down are from various regions across Indonesia; they also have diverse living traditions and culture. Thus, it is difficult to unite the mindset of newcomers with that of the natives and migrants who have been living in Jakarta for a long time. It is expected that the mindset of this diverse population is set to care about the environment, so that Jakarta is spared from catastrophic floods that often take place every year. The increasing level of urbanization that aims at finding employment for better life or economic status is not achieved but turns into suffering, since it is difficult to obtain employment, and the result is poverty. This has incurred even more problems to the local government. In these circumstances, the level of public awareness of the environment will be reduced or low. Thus, the flood is essentially created by the people themselves. So far the government has dealt with floods by helping communities hit by disasters, manage waste disposal, asking people in flood-prone location to move to other places, as well as other efforts. The risk of floods becomes more severe due to the high population density. The population of Jakarta at night is 8,522,545 or more than 10 million people during the day due to the arrival of workers from outside the city. East Jakarta is the most populous part of the city with 2,626,708 people (see Table 1).

Table 1: Population and area of DKI Jakarta: month: February 2011.

Region	Indonesian Citizens			Foreign Citizens			Total
	Male	Female	Sum	Male	Female	Sum	
Central Jakarta	503,411	418,822	922,233	191	146	337	922,570
North Jakarta	776,582	644,973	1,421,555	269	240	509	1,422,064
West Jakarta	869,057	765,332	1,634,389	334	302	636	1,635,025
South Jakarta	1,062,260	831,361	1,893,621	408	251	659	1,894,280
East Jakarta	1,426,586	1,199,896	2,626,482	121	105	226	2,626,708
Thousand islands	11,438	10,460	21,898	0	0	0	21,898
Total	4,649,334	3,870,844	8,520,178	1,323	1,044	2.367	8,522,545

Source: Office of Population and Civil Registration City Administration [2].

Jakarta with a highly-dense population – an average of 12,835 people in each square kilometre – greatly disturbs development efforts that aim at establishing a prosperous society. In reality, in flood-prone areas, residents face an increasingly difficult life. Yearly occurrence of floods should be a warning to the people of Jakarta for caring and being aware of the environment. According to the theory, a form of awareness is expressed through efforts to reduce floods by: cleaning up sewage/drainage system from mud sediment, proper disposal of waste, and other social activities [3]. The form of awareness must be great since floods take place frequently; thus, public environmental awareness should be increasing.

2 Discussion of research

Discussion on the research is important as it will analyze the reason why the community of Jakarta, who are often affected by floods, prefer to stay in their respective places in spite of the government's frequent encouragement for them to move to other safer places.

2.1 Purpose of the research

This research aims at determining the level of environmental awareness knowledge of the people of Jakarta, the level of flood risk knowledge, and the behaviour of people in flood-prone areas of the environment, which affect the level of flooding risk in Jakarta. The study was conducted in September–November 2010. The risk of flooding in Jakarta is very saddening and it is necessary to find out the actual reaction of the affected community.

2.2 Methodology

Survey and field observations have been employed in data collection for the research. A sample of the research was East Jakarta with respondents of the same behaviour characters as those of other regions. East Jakarta comprises quite a vast area and huge population. In addition, the distribution of flood-prone area in this region is widely spread. Forty families, with parents and their respective sons and daughters in each family, have been selected as samples of the study by means of purposive sampling. This made a total number of 160 respondents involved in the study. Research instruments consist of questionnaires and observation protocol. A blueprint of the questionnaire has been set up with reference to the environmental awareness indicators and sub-indicators derived from related theoretical framework, i.e., environmental knowledge and behaviour/attitude on the environment. Calibration of the instrument was performed prior to data collection (two invalid items deleted from an initial 17 items, leaving 15 items for use in the questionnaire) with instrument reliability at (r) = 0.81. Simple qualitative statistical analysis (matrix and percentage) has been employed in processing the obtained data.

2.3 Results of the research

2.3.1 Description of data

Based on survey results, 2009/2010 data of the Flood Control and Evacuation Management Task Force Unit of the East Jakarta Local Government suggests ten flood-prone districts in the region. More information can be observed in the data presented in Table 2.

Table 2: Flood-prone districts and sub-districts in East Jakarta region.

No.	Flood-Prone Districts	Sub-district/ Population	Sub-district/ sample	Remarks
1	Pulogadung	4	Five sub-districts selected by random sampling: Pekayon, Kebon Manggis, Kayu Putih, Cipinang Muara, and Lubang Buaya	Twenty-seven flood-prone sub-districts were identified in ten existing districts
2	Matraman	1		
3	Jatinegara	5		
4	Kramat Jati	4		
5	Pasar Rebo	1		
6	Cakung	4		
7	Duren Sawit	3		
8	Makasar	3		
9	Cipayung	1		
10	Ciracas	1		

2.3.2 Residents' perception

Data was collected by means of questionnaire and observation protocol from flood-affected residents. Questionnaires were completed by 160 respondents, who came from 40 families: 40 father respondents, 40 mother respondents, 40 son respondents and 40 daughter respondents. Field data was then tabulated and summarized with reference to the set out indicators; i.e. knowledge and behaviour/attitude on environment.

2.3.2.1 Environmental awareness knowledge

Knowledge is everything we have in our understanding that comes from practice. Practice produces experience and experience produces knowledge. Knowledge of environmental awareness is the experience of caring for the environment, cleaning up sewage/drainage system, disposing of waste properly, etc. According to Husserl cited in Brauwer, awareness is the conscious mind or knowledge that controls reasoning, conscious living, part of attitude and behaviour, and is described as a phenomenon of nature that must be explained based on causality principles [4]. Joseph Murphy, consciousness is one's state of awakeness or awareness of his/her own behaviour, i.e. the conscious mind that controls the mind and determine one's desired choices/preference such as good or bad [5]. Controlling flood in Jakarta is not enough just by calling a warning to be wary of potential floods; more active action is necessary to eliminate major problems that trigger the floods. As a matter of fact, in the case of floods, lack of environmentally aware behaviour has been the root problem. Thus, such behaviour needs be changed into environmentally aware behaviour. To be clearer, data related to environmental awareness knowledge is presented in Table 3.

Indicators of the level of environmental awareness knowledge of Jakarta people presented in Table 3 are of behaviour; they refer to one's behaviour on his/her immediate environment that are results of his/her knowledge. The indicators, among others, are maintaining cleanliness of the environment, care

Table 3: Level of environmental awareness knowledge of Jakarta community.

No.	Statements	Percentage (%)				High Percentage (%)
		Father	Mother	Son	daughter	
1.	Maintaining cleanliness of the environment	32.50	10.00	5.00	12.50	15.00
2.	Caring for the environment	15.00	30.00	25.00	17.50	21.88
3.	Protecting the environment	17.50	10.00	5.00	12.50	11.25
4.	Maintaining the environment	15.00	22.50	10.00	15.00	15.63
5.	Attempting environment conservation	7.50	5.00	7.50	7.50	6.87
6.	Disposing of waste properly	5.00	7.50	17.50	7.50	9.38
7.	Cleaning up the environment	2.50	2.50	10.00	5.00	5.00
8.	Keeping a natural living place	5.00	7.50	5.00	5.00	5.62
9.	Performing the green revolution/reforestation	-	2.50	2.50	7.50	4.16
10.	Living a healthy and clean life; Keeping a love for the environment	-	2.50	12.50	10.00	8.33

for their environment, love of the environment, dispose of trash in its place, reforestation activities and so on. As presented in Table 3, there are indicator statements with higher percentage score than others. The scores reflect the respondents' level of environmental awareness knowledge. Each of the indicator statement items represents the level of environmental awareness knowledge of the community. The indicator that obtains the highest score is the *Caring for the Environment* indicator, which determines the level of environmental awareness knowledge of the community. Therefore, the level of environmental awareness knowledge of the Jakarta community, according to analysis results, is 21.88%.

2.3.2.2 Knowledge of flood risks

When the rainy season arrives, some regions in Indonesia, especially Jakarta floods always take place. In 2010 and early 2011, Indonesia received abundant rain fall, and again there was flood in every province. Floods that occurred across Indonesia are caused by overflowing rivers. Regions where floods often occur are those stretched out along the rivers. To anticipate the catastrophic floods that often occur, it is necessary to clean up or dredge the river bed (there are sixteen rivers in Jakarta). Dredging activities attempted by the government are quite positive. Yet, since the root of problems is not addressed, then, a month later, the rivers will be full again with rubbish. Due to high rainfall, rivers in Jakarta overflow and floods occur. The potential risk of floods is enormous, including the potential loss suffered by the industry located in close-to-river regions. Areas that should have served as a water shed, in the rainy season, turn into flood supplier. On the contrary, in the dry season, the river flow is reduced drastically so that it is hot and dry in the city.

Today attempts to control floods in Jakarta have changed in terms of paradigm and actions. Disaster management is focused more on community participation in the management. Thus, the community are not merely victims or objects of the disaster, but also as disaster management actors. This community-based program encourages or motivates community capacity empowerment to be alert in preventing and eliminating risks and the impact of floods that occur in their immediate living environment. The growth of development in Jakarta is going fast, yet, lack of attention for the city layout structure, which only contributes towards flood risks. Furthermore, the very little amount of open, green space has led to a lack of water catchment areas that can also cause floods. Therefore, when rain comes, even a large water canal will quickly overflow. Lack of maintenance of the drainage systems in several locations has been the cause of floods. Findings of field surveys and observations provide evidence that citizens who are often exposed to floods have very poor knowledge about floods. Data on the knowledge of flood-victim respondents is presented in table 4.

The Jakarta local government is aware that there is a lack of green open space as a water shed; therefore, the Governor plans to expand the green open space area in Jakarta. The city is about 650 km^2. According to legislations, the ideal green open space is 30% of the total area. It now has only 9.8% [6]. Data on the level of knowledge of flood risks presented in Table 4 shows several indicators that demonstrate the society's level of knowledge of flood disaster. The indicator

Table 4: Level of knowledge of flood.

No.	Statements	Percentage (%)				High Percentage (%)
		Father	Mother	Son	Daughter	
1.	Rains in the upland, overflowed rivers upland, flood occurs in lowland.	42.50	50.00	40.00	40.00	43.12
2.	Flood originates from a higher location	17.50	25.00	5.00	-	15.83
3.	Flood has been due to rain water supply from Bogor	15.00	10.00	22.50	-	15.83
4.	Flood originates from other higher place/land	10.00	7.50	5.00	22.50	11.25
5.	Continuous rain in Bogor has incurred flood	7.50	7.50	12.50	7.50	8.75
6.	Flood was due to overflowing rivers in Bogor	5.00	-	10.00	-	7.50
7.	Flow of rain water from Bogor cannot be blocked	2.50	-	-	10.00	6.25

with a high percentage score is a flood coming from upland areas. This indicator suggests that when there is heavy rain upland, flood will take place in low areas. Controlling the flood can be done by means of concrete actions; i.e. by training people to not only be sufferers in the disaster but act as actors in disaster management as well. The highest score is obtained by the ‘upland area’ indicator. Therefore, there is a need to find out their level of knowledge about flood due to the fact that flood can occur at any time. In Table 4, indicators of knowledge about flood risks can be observed. Each indicator item represents the level of knowledge about flood risks of the Jakarta people. Thus, with reference to data analysis results, the level of knowledge about flood risks of the Jakarta people is 43.12%.

2.3.2.3 Flood-affected citizens refuse to move out of the flood-prone areas

As a matter of fact, the Jakarta local government is actually so experienced in handling flood risks that they can minimize gaps and weaknesses. Jakarta citizens, particularly those who are affected by floods and live by the river banks, understand well that their living area is subject to recurring floods. They are not allowed to remain in the flood-prone areas or by the river banks. Yet, this is difficult to implement since they are not willing to be relocated or move to other residential areas. To be clearer, Table 5 presents data on flood victims who are not willing to move from flood-prone areas.

Data presented in Table 5 shows the proportion of flood victims who refuse to move from flood-prone areas. They become a burden to the local government who has been responsible for taking good care of them until now. On studying the indicators that represent reasons for residents refusing to move from their current living area, some surprising things can be observed. Their reasons

Table 5: Percentage of flood victims who refuse to move from flood-prone areas.

No	Statements	Percentage (%)				High Percentage (%)
		Father	Mother	Son	Daughter	
1.	Have no funding support	42.50	32.50	-	-	18.75
2.	Already feels comfortable living in current place	27.50	42.50	50.00	55.00	43.75
3.	Have not yet found a better place to live	15.00	12.50	-	-	6.88
4.	Current flood-prone area is a very strategic place to live	12.50	12.50	17.50	20.00	15.62
5.	Current house brings great luck	2.50	-	-	-	0.63
6.	Currently living with parents	-	-	32.50	25.00	14.38

includes *no funding support for moving to other place* and *Already feels comfortable living in current place.* This is indeed surprising. Despite their being threatened by risks of flood, they feel comfortable, happy and safe, living in that place. Based on data in Table 5, the percentage of people who are not willing to be relocated from flood-prone areas for they do not have sufficient funding to do so is quite a lot too. There are also residents who are not willing to be relocated because they still live with their parents who live in the flood prone areas. Some others remain to live in the flood-prone areas because the location is strategic and brings good luck to them. Yet, the most surprising indicator is the fact that they are not willing to move because they feel comfortable living there. This indicator obtains a percentage score of 43.75%. These findings can be considered true as relocating residence from flood-prone areas is quite difficult and becomes a great burden to Jakarta local government.

2.3.2.4 Community's attitude on environment

It has been explained that the Jakarta local government's efforts to manage the risks of yearly flood is quite great. This is because the disaster management efforts experience a change in terms of paradigm and concrete actions. Disaster management has been focused on community participation in managing the flood disaster. This community based program encourages community capacity empowerment to be alert for preventing and eliminating the accompanying impact and risks on the immediate environment. This program has been implemented for the community as affected subjects need to be empowered by providing them with the necessary knowledge and skills that will enable them to manage impact of the disaster and eliminate the risks of flood. The Jakarta local government has attempted maximum efforts in controlling flood risks. However, the researcher observes that they have not yet fully addressed the fundamental aspects, i.e. attitude and behaviour, which is important in order to reset the existing mindset of the people to a new mindset that is acceptable to the whole community. Michael O'Driscoll states that issues related to changes in attitude also involve a self-motivation in attempting the change. Efforts to encourage a change in attitude usually occur following exposure to new or a different piece of information about the object or problem in question [7].

The above statement signifies that a change in attitude also requires modification prior to establishment of the newly acquired belief. Such way of thinking brings differences between two major strategies for changing attitudes; that is, active participation and persuasive communication. Besides, compliance occurs when one receives the influence of other(s) for social acceptance reasons. As an approach, active participation strategy requires active involvement of individuals, which is determined by the individuals' contact with the government who functions as initiator. Survey and observation reveal data on the community's attitude to their immediate living environment related to risks of flood. An indicator of the community's attitude through concrete actions on the flood-prone environment can be observed in Table 6.

One indicator statement in the table shows quite a low level of environment awareness knowledge; i.e. does not feel sad or disappointed on seeing illegal constructions alongside the river (28.75%). However, when the average score of indicators of behaviour is calculated, the result comes to 47.25%. This indicates that people living in the flood-prone areas are not yet performing an environmentally friendly behaviour. This description signifies that human beings should make friends with their immediate living environment; else, they will be disadvantaged. Chiras emphasizes that men are part of nature. They are not superior to nature. They are dependent on their immediate living environment that they need to always think of taking care of nature and cannot live without [8].

In reality, this is not the case. People seem to be unfriendly to their environment. Their daily conversations seem to suggest that they get along well with their environment, yet, it is not reflected in their actions. This goes on and shows that the Jakarta community remains less friendly with their environment or shows a lack of environmental awareness, particularly when it comes to taking care of the environment in flood-prone areas. Similarly, the drainage system in Jakarta, which is mostly of a closed type, cannot be expected to function properly as it is full of garbage/rubbish and in need of cleaning/dredging. Indeed, this may incur great cost. Furthermore, the local government has not yet shown firmness in promoting discipline and implementing regulations on environmental issues [9]. They fail in issuing warnings or applying penalties to people who dispose of waste carelessly. Therefore, the core problem in environmental awareness issues lies in the behaviour of the people or the community. Regulations are made to be obeyed and implemented firmly by all members of the community including the leaders. In other words, environmental awareness, particularly the flood disaster management efforts, can be established when all members of the community are involved, i.e. the leaders and followers/citizens. Leaders are expected to be models in caring for environment activities. Particularly in Jakarta, they need to be firm in building discipline related to environmental care. It is necessary to apply a sanction for improper waste disposal practices with local government regulations as reference [10]. Existing regulations should be applied properly and firmly. Causes of yearly floods and traffic jams in Jakarta are our own practices, human actions, and we should not blame each other or nature for the problems earned.

Table 6: Data on community behaviour on the environment in flood-prone location.

No.	Statements	Percentage (%)			High Percentage (%)
		1	2	3	
1.	Cleaning the drainage in home front	16.80	42.50	37.50	42.50
2.	Given the task of sweeping garbage	5.00	35.00	60.00	60.00
3.	Social service work with parents	17.50	41.25	41.20	41.25
4.	Planting trees in the yard	30.62	39.30	30.00	39.30
5.	Make an absorption hole or bio-pore	25.62	33.75	40.62	40.62
6.	Sad when flooding	25.00	36.87	51.25	51.25
7.	Not sad to see illegal buildings along side of the river	11.87	19.75	28.75	28.75
8.	Setting up the levee board retaining flood	47.50	23.62	50.00	50.00
9.	Processing of organic waste becoming fertilizer	26.62	39.38	25.00	39.38
10.	Creating high homes	34.38	13.12	79.40	79.40
Average Percentage					**47.25%**

According to Engel and Engel, proper development must correspond with high level human activities; that is, in obtaining the values of life, man must live properly on earth with regard to environmental ethics. It is because the earth is theirs, but not exclusive to them. Furthermore, the end of a development should be a better state [11]. For this reason, it is necessary for development activities in Jakarta to consider carefully existing physical environment, climate, and aspects related to city plan and lay-out. In other words, it sufficiently attends to availability of green open space and non-green one to function as a watershed for the city as well as improving the poor drainage system. It is recommended that the function of wetlands that have been closed for housing development be restored, reduce the use of building materials or paving materials in the form of asphalt, concrete and cement that are not water absorbent, and replace them with more environmental friendly materials. Besides, water is one source of life that should be utilized wisely instead of turning it into disaster for people. Eventually, it is realized that the risk of floods that occurs annually has been the result of us not taking advantage of the rain as God's blessing. In fact, the rainy season brings about terrible disaster to people. On the contrary, in the dry season, flood-prone areas experience water shortage or drought. Such a situation signifies the inability of human beings in taking best advantage of God's abundant blessing. Therefore, it is necessary to learn from nature that God provides us with.

3 Conclusion

In conclusion, the level of environmental awareness knowledge of community in Jakarta is 21.88%; level of knowledge about risk of floods is 43.12%, people's behaviour on the environment in locations vulnerable to floods is 47.25%, and flood victims who are not willing to move from flood-prone areas is 43.7%. The low level of awareness has been influenced by the diverse cultures, science and technology, and globalization. Competition in life, which is quite high, hard, and difficult, has left people with no time to think about caring for the environment. It is expected that the community demonstrate positive behaviour by practising environmentally aware actions such as the proper disposal of waste/trash, cleaning up the drainage system from trash, and supporting the government's flood control and risk management programs. All efforts attempted by Jakarta local government can be considered optimum and need to be supported by the whole community.

Attitude and behaviour aspects are so important that they should be a concern to us. There is a need to change the mindset of the society into one that cares about the environment. Jakarta local government needs to change or improve its method of communication into a participatory method of communication through harmonious and persuasive participation with the people in the flood-prone areas. This needs to be done with a view to helping people living in flood-prone areas to understand the threatening danger occurring every year and make them willing to move to safer and more comfortable living areas. Therefore, it is important for development efforts to be guided by existing city layout plans; that aim at improving current poor drainage system management and, in general, controlling the abundant water supply in the rainy season, and making the best of it to support prosperity of the people rather than turning it into a catastrophe.

References

[1] Sajogyo, Pudjiwati. *Development Sociology*. Jakarta: FPS IKIP Jakarta and BKKBN Jakarta, pp.119-128, 1985.

[2] Office of Population and Civil Registration City Administration, 2010

[3] Neolaka, Amos. *Environmental Awareness*. Jakarta: Rineka Cipta, pp. 79-86, 2008.

[4] Husserl who quoted Brauwer, M.A.W. *Human Body in the Light of Psychology Phenomenologist*. Jakarta: Gramedia, pp.56-65, 1986.

[5] Murphy, Joseph (1988). *Generating Power of the Subconscious*. Bandung: Pionir Jaya CV, pp. 19-25, 1988.

[6] National Disaster Management Agency. *Disaster Management Floods* Jakarta: Public relations. Jakarta, 29-11-2010.

[7] Michael O'Driscoll, Godfrey Gardner, J Michael Innes, Joseph P. Forgas Philip L Pearce, James W. Newton. (1981). *Social Psychology*. Sydney: Prentice-Hall of Australia Pty Ltd, pp. 294-301. 1981.

[8] Chiras, Daniel D (1991). *Environment Science Action for a Sustainable Future*. California The Benjamin/Cumming Publ. Company Inc. Third. Ed., pp. 454-464, 1991.
[9] RI Law No.32 on the *Protection and Management Environment*, 2009
[10] Government Regulation No15 About *Implementation of spatial*, 2010
[11] Engel, Ronald .J & Engel, Joan Gibb (1990). *Ethics of Environment and Development, Global Challenge and International Response*. London: Belhaven Press, pp. 27-40, 1990.

Section 4
Flood forecasting

A multivariate model for flood forecasting of lake levels

M. Mohssen
Department of Environmental Management,
Lincoln University, New Zealand

Abstract

A new multivariate flood forecast model for lake levels has been developed. The model is based on the concept of the projection theorem, to obtain the projection of lakes flood levels on total rainfall forecasted or observed at nearby rainfall gauges. A comprehensive analysis of 20 historical and significant rainfall events on Lake Wakatipu and Lake Wanaka in the South Island of New Zealand has been carried out to detect the best time series and functions to be used for this projection.

The study showed that linear relationships exist between total lake rise and total rainfalls observed at available rain gauges. In addition, during flood events, linear relationship can be utilised to represent the function between observed cumulative rainfall and cumulative lake rise at 10 hours (Lake Wakatipu) and 7 hours (Lake Wanaka) of lead time. Analysis of relationships between lake rise and alternative options for available rainfall sites resulted in the choice of rainfall sites for the best forecast model. Parameter estimation of a multivariate model which includes both Lake Wakatipu and Lake Wanaka flood levels and the selected rainfall sites was carried out by using the projection equations of orthonormal sets. The derived model was applied to two observed events which were not included in the calibration process. The new model has performed well in forecasting flood levels of both lakes.
Keywords: flood forecast, flood modelling, rainfall-runoff, lake level, projection theorem, Hilbert space.

www.witpress.com, ISSN 1743-3541 (on-line)
doi:10.2495/FRIAR120101

1 Introduction

Floods are the most common natural disasters, and they cause devastating damage to communities. In the absence of flood protection infrastructures, flood warning is one of the most effective schemes to mitigate the impact of these natural hazards. The Environment Agency of UK and the strategic plan for the US National Weather Service indicated the urgent need for major investment to develop new forecast models for flood warning [1, 2].

A reliable flood forecast model is the most important component of an efficient flood warning system. If reliable estimates of flood levels are forecasted before the event happens, authorities/communities can have sufficient time to mitigate the impact of the coming flood event by getting prepared, evacuating, moving stock away from flood prone areas, and relocating precious items. Despite the fact that literature is rich with research on flood forecasting, many of these models/techniques fail to accurately forecast flood events on the shorter scale (hourly) due to the high variability of the associated hydrological and meteorological processes [1]. High variability of precipitation, the main driving factor for floods, along with complex hydrological characteristic of the catchment area, makes it hard to obtain the right variables representing these processes at the right resolution to be capable of producing a reliable flood estimate of an occurring or incoming event.

Despite the fact that the hydrologic process of lake flooding is quite different from river floods, available modelling approaches in the literature are still similar. Watershed modelling is used to simulate the inflows to the lake, and then a hydrologic budget model is applied to translate the difference between inflows and outflows to a rise (or drop) in lake level [3, 4].

Time series modelling, by using ARIMA models have been applied in the literature, but due to the stationary requirement of the modelling process, usually they apply to longer time periods (days or more), while during flood events we are usually concerned in hourly flows [5]. ANN has been recently applied to forecast lake levels, and several techniques have been suggested for their applications to hourly time steps [6–8].

Floods are the most damaging and costly natural hazard in New Zealand, with about 935 damaging flood events occurred during the period 1920 to 1983 [9]. Queenstown, a major town by Lake Wakatipu which is full of attractions for tourism, has been badly flooded during the events of 1878 and 1999. Flooding of Lake Wakatipu takes a long time (could be weeks) to recede, and this will have persistent impact on the economic activity of Queenstown township and its community. Lake Wanaka, about 60 km north east of Queenstown has also flooded the township of Wanaka which is on the lake front.

2 Multivariate forecast model for lake levels

Lake Wakatipu has an area of about 293 km^2, with a total catchment area of about 3059 km^2, and lies in the western ranges of Otago district, in the South Island of New Zealand. Lake Wanaka has an area of about 196 km^2 and its catchment is also in the western ranges of Otago with an area of 2564 km^2. Lake Wanaka discharges its outflows to the head of the Clutha River, which is the largest river in New Zealand with average flows of 534 m^3/s, while Lake Wakatipu outflows to the Kawarau River, which joins the Clutha River further downstream. Figure 1 shows the two lakes, their outflows and the location of rainfall sites of this study.

Four rainfall sites (the Hillocks, Makarora, Peats Hut and West Wanaka) are included in this study, as they have a suitable long record of data, while other sites are shorter and they don't include the most significant flood event of November 1999. Lake Wakatipu outflows are recorded at Willow Place, while the site for Lake Wanaka outflows is at Roys Bay, as shown in Fig. 1. The third lake "Lake Hawea" which lies just to the east of Lake Wanaka was not included in this study as it is completely controlled, and there is no major settlement beside the lake.

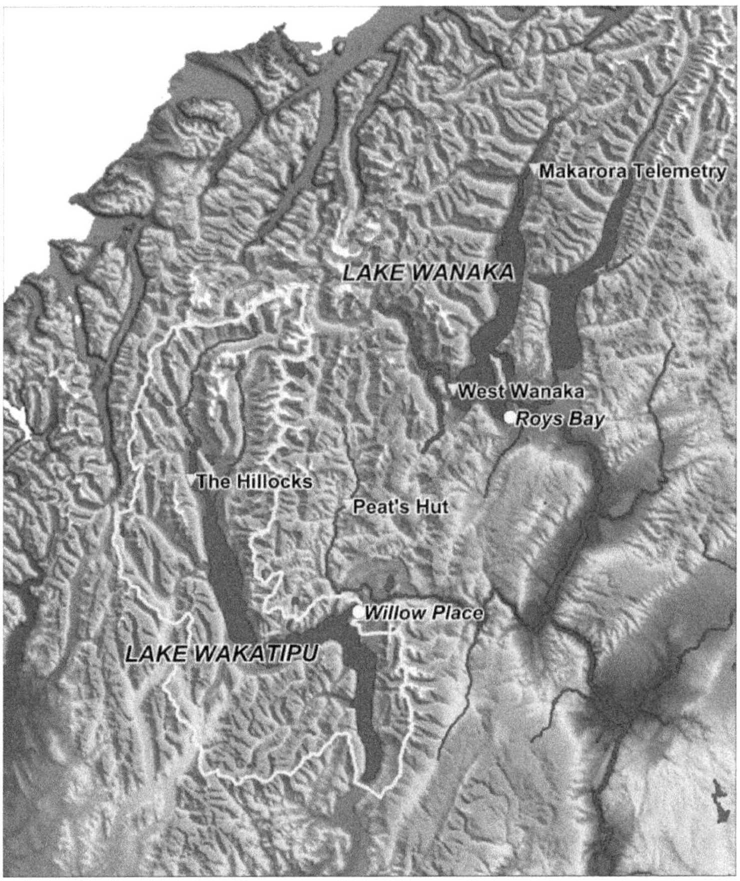

Figure 1: Rainfall and flow sites of Lake Wakatipu and Lake Wanaka.

Lake flooding is different from river flooding as the lake has a huge storage and its outflows are a function of its level rather than inflows to the lake. Thus lake rise is a function of cumulative rainfall over a period of time rather than its intensity during a short period of the rainfall event [10]. The new concept for modelling which has been applied in this research utilises the projection theorem [11] to obtain the projection of the total rise of lake levels on the cumulative rainfall of the selected rain gauges. The projection theorem guarantees that the obtained model will produce the best forecasted flood levels, if the correct function which relates the time series was selected. The work carried out here is an extension of the univariate case presented by [10], where Lake Wakatipu levels were estimated by rainfall at only one rainfall gauging site (the Hillocks). In this research, multiple rainfall sites will be used to forecast lake levels at two lakes (Wakatipu and Wanaka).

2.1 Model formulation

For large lakes, such as Lake Wakatipu and Lake Wanaka, flooding will be slow and flood levels are more dependent on cumulative rain rather than rainfall intensity, and usually peak flood levels recede slowly as the lake outlet takes long time to dispose the huge amounts of water stored in the lake during the flood event. An hourly hydrologic balance for a lake, in the univariate case, can be expressed as follows [10]:

$$L_{t+t_f} = L_{t-t_1} - \sum_{j=t-t_1}^{j=t+t_f-1} Q_j + f(R_t) \tag{1}$$

Assuming that at the start of the rainfall event t=0, then L_t is the lake level (mm) at present time t, t_f is the forecast time in hours after t, t_1 is a lag time (hrs) before t, Q_j is the lake outflow (mm) at time j, R_t is the cumulative rainfall (mm) over a period of hours and $f(R_t)$ is a function of this cumulative rainfall. Of course $f(R_t)$ will have units of level (mm).

Mohssen and Goldsmith [10], by analysing the 1999 flood event, showed that the best value for t1 is the present time t, thus $t-t_1=0$, which is the start of the rainfall event. Level L_0 at the start of the event is usually used to get the rise of the lake level. Thus, Eq. (1) can be written as follows:

$$L_{t+t_f} - L_0 + \sum_{j=0}^{j=t+t_f-1} Q_j = f(R_t) \tag{2}$$

Let us denote the forecasted total lake rise since the start of the event $L_{t+t_f} - L_0$ as ΔL, and the cumulative outflows since the start of the event $\sum_{j=0}^{j=t+t_f-1} Q_j$ as ΔQ, then Eq. (2) in the multivariate form is:

$$\begin{bmatrix} \Delta L_1 \\ \Delta L_2 \end{bmatrix} = \begin{bmatrix} f(R)_1 \\ f(R)_2 \end{bmatrix} - \begin{bmatrix} \Delta Q_1 \\ \Delta Q_2 \end{bmatrix} \tag{3}$$

where the subscripts 1 and 2 refer to Lake Wakatipu and Lake Wanaka, respectively. Total lake inflows during a flood event equal the lake rise plus cumulative outflows, i.e. $\Delta\mathbf{I} = \Delta\mathbf{L} + \Delta\mathbf{Q}$, thus:

$$\Delta\mathbf{I} = \mathbf{f}(\mathbf{R}_t) \tag{4}$$

where $\Delta\mathbf{I}$ is a column vector of total lake inflows during the flood event at the two sites, $\mathbf{f}(\mathbf{R}_t)$ is a column vector of functions of rainfall at multi rainfall sites. Mohssen and Goldsmith [10] showed that using total lake inflows does not result in improving the reliability of the forecast model for the univariate case, and to the contrary, model testing showed that it resulted in less reliable models. The same has been shown in the results of this study for multi rainfall sites, and thus we will show only the results for the projection of total lake rise since the start of the event on cumulative rainfalls at selected sites.

For the best projection of $\Delta\mathbf{L}$ on $\mathbf{f}(\mathbf{R}_t)$, $\Delta\mathbf{L} - \mathbf{f}(\mathbf{R}_t)$ should be orthogonal to $\mathbf{f}(\mathbf{R}_t)$. In Hilbert space, this is written as:

$$<\Delta\mathbf{L} - \mathbf{f}(\mathbf{R}_t) , \mathbf{f}(\mathbf{R}_t) > = 0 \tag{5}$$

Assuming that:

$$\mathbf{f}(\mathbf{R}_t) = \propto \boldsymbol{X} = \textstyle\sum_1^n \propto_i \boldsymbol{X}_i \tag{6}$$

where α is a row vector of parameters, and **X** is a column vector of power functions of rainfall at each site R_i, Eq. 5 can be re-written as:

$$<\Delta\mathbf{L} - \propto \boldsymbol{X} , \boldsymbol{X} > = 0 \tag{7}$$

Application of Eq. (7) would produce n equations in the parameters ∝, which can be solved simultaneously to obtain the values of the parameters ∝ as a function of statistical properties of the time series ΔL and X.

The solution of (7) can be written, in matrix form, as follows:

$$\propto = (\boldsymbol{X}^T\boldsymbol{X})^{-1}\,\boldsymbol{X}^T\,\Delta\boldsymbol{L} \tag{8}$$

In case $\left(\boldsymbol{X}^T\boldsymbol{X}\right)^{-1}$ is singular, there will be infinite solutions of (8). However, by the uniqueness of the projection theorem, all of them will produce the same forecast [11].

The projection theorem guarantees that the model provided by (8) will produce coefficients of $\mathbf{f}(\mathbf{R}_t)$ for the best forecasts of $\Delta\mathbf{L}$. Of course, the type of the functions $\mathbf{f}(\mathbf{R}_t)$ is not known, and part of the analysis carried out here is to obtain the best estimate of this function. In the case of a linear relationship, the power of R is one, and the power can be zero for a constant parameter (intercept). There are two different cases to be considered herein. The first case is the total lake rise (inflow) due to the total rainfall of an event. This is required in the case of long term forecast, when an event is expected to hit the region with an estimated total rainfall at a selected site, for instance on the second day or two days later. In this case, real rainfall of this event is not observed yet, but it is important to forecast the impact of this event on the potential flooding of Queenstown or Wanaka. The other case is flood forecasting during the event itself, and in this case one forecasts the cumulative lake rise at both lakes after several hours (forecast time) due to observed cumulative rainfall at the selected rainfall sites.

2.2 Case 1: Total lake rise due to total rainfall

Figures 2 and 3 present the relation between total rainfall of 20 events, and the corresponding total lake rise. The figures show that a linear relationship is quite suitable to represent this relationship. It is quite clear from Fig. 2 that total rise of Lake Wakatipu is represented well by a linear relation of total rainfalls at the

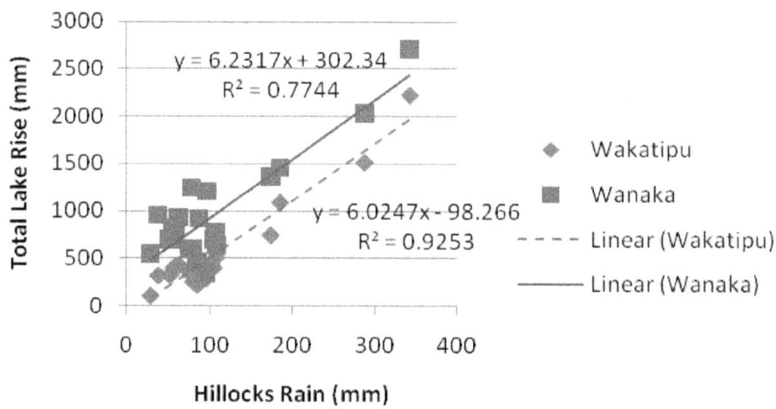

Figure 2: Total rainfall at Hillocks and Lake Wakatipu Rise.

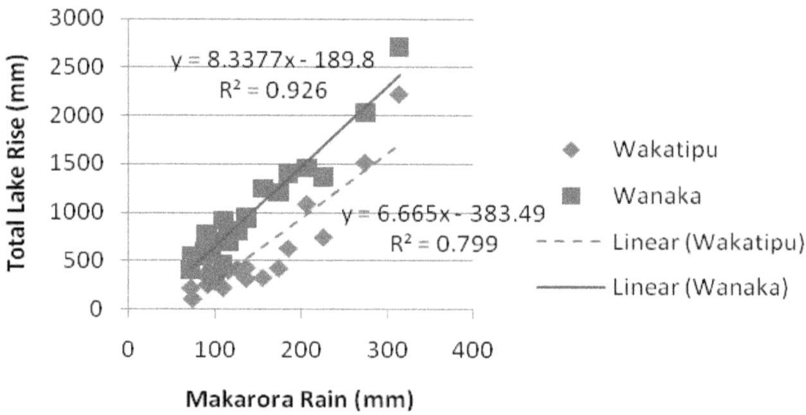

Figure 3: Total rainfall at Makarora and Lake Wanaka Rise.

Hillocks (R^2 = 0.93), while this is not the case for Lake Wanaka. Makarora rainfall site offers a significantly improved linear relationship to the total rise of Lake Wanaka (R^2 = 0.93) compared to the Hillocks rainfall (R^2 = 0.8). These results indicate the importance of carrying out the analysis to choose the right rainfall sites to select for the modelling process. Figures 4 and 5 show the linear relationship for the multivariate case. Both lakes have slightly improved their relationship with rainfall at the selected sites. However, the figures again confirm the fact that the wrong choice of rainfall sites would result in an unreliable forecast model. It should be mentioned that, in this study, other alternatives for rainfall sites were included, but only cases which improved the linearity of the relationship are presented.

Note that the variable x of the trend line in Figs 4 and 5 is the weighted average of total rain at several rainfall sites according to the weights of the projection (multiple regression) of lake rise onto total rainfall of these sites. Moreover, the projected weighted coefficients of the selected rainfall sites are not the same values for both lakes. For instance, in Fig. 4, the shown graphs are for the fitted parameters for Lake Wakatipu, not Lake Wanaka, and that is why it shows much higher determination coefficient for Lake Wakatipu than for Lake Wanaka, and the opposite is true for Fig. 5. Cases where multiple regression produced negative coefficients for any of the rainfall sites were excluded, as one would consider these coefficient to represent the contribution of this rainfall site to the lake under study. The models derived from this study are as follows:

$$\begin{bmatrix} \Delta L\ (Wakatipu) \\ \Delta L\ (Wanaka) \end{bmatrix} = \begin{bmatrix} 2.54 & 1.42 & 3.72 & -165 \\ 0.00 & 7.13 & 1.69 & -134 \end{bmatrix} \begin{bmatrix} H \\ M \\ W \\ 1 \end{bmatrix} \tag{9}$$

where total rainfall of the event at the Hillocks is H, at Makarora is M and at West Wanaka is W.

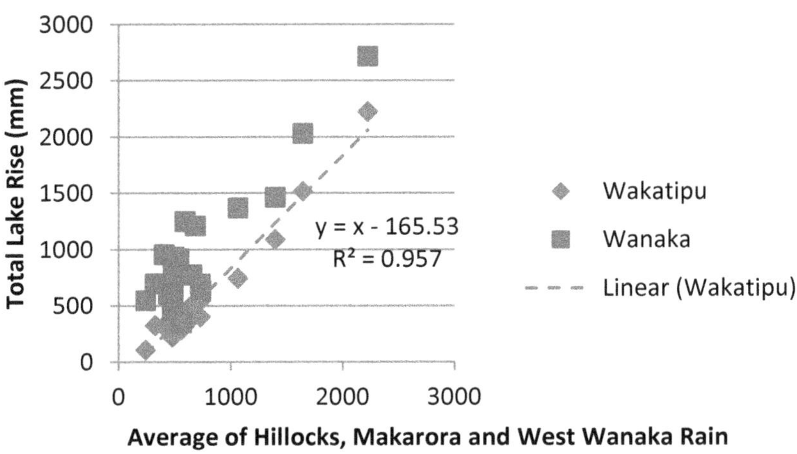

Figure 4: Average weighted rainfalls and Wakatipu level rise.

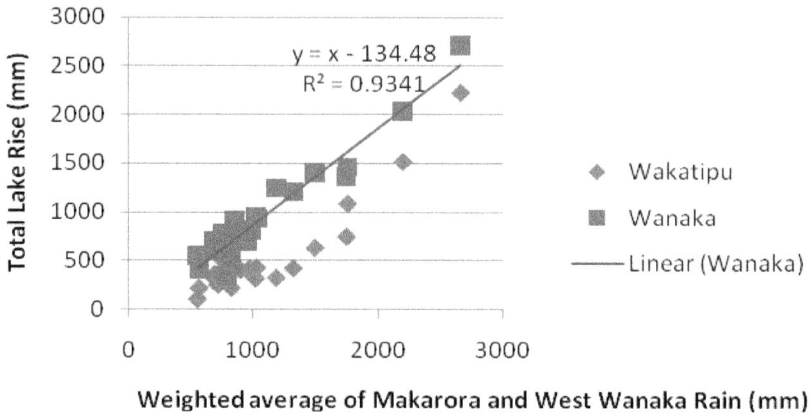

Figure 5: Average weighted rainfalls and Wanaka level rise.

2.3 Case 2: forecast during the rainfall event

Mohssen and Goldsmith [10] showed that a lag of 10 hours produces a good linear relationship between cumulative Lake Wakatipu rise and cumulative rainfall at the Hillocks since the start of the rainfall events. As shown in Fig. 6, the optimum correlation between cumulative increase in Lake Wanaka rise and cumulative rainfall since the start of the rainfall event is 7 hours. Figure 7 confirms that linear relationship exists between cumulative rainfall and cumulative Lake Wanaka rise, and that the use of lag 7 hours significantly improved this linear relation, which will definitely improve any modelling process. The results of Fig 7 show the effectiveness of this new technique in improving the derived model.

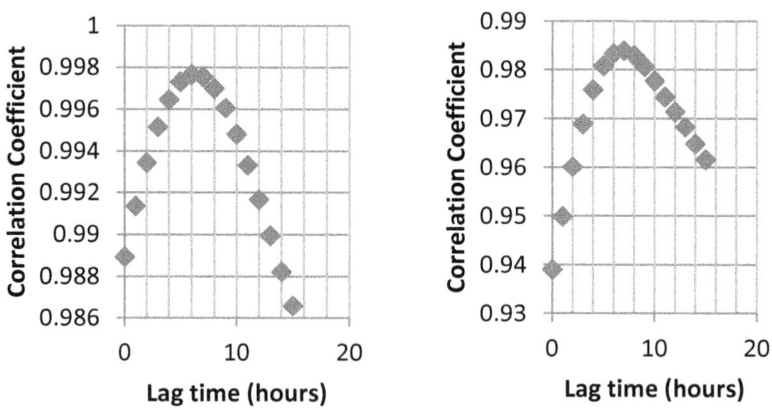

(a) November 1999 Event (b) April 2010 Event

Figure 6: Lagged correlations between cumulative rain and Lake Wanaka rise.

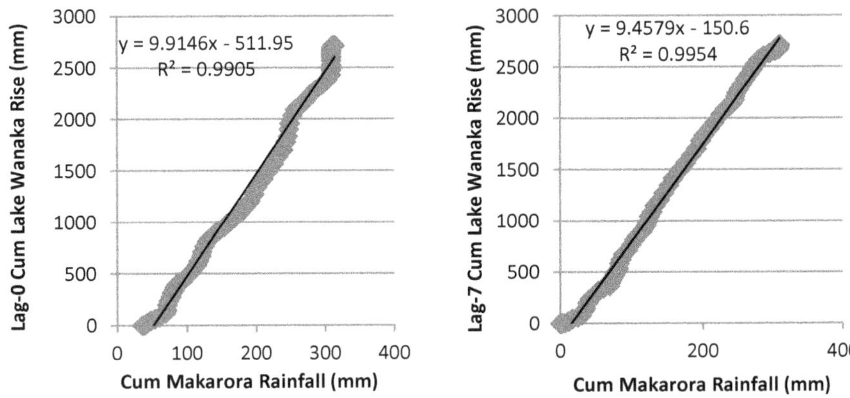

(a) Lag-0 (b) Lag-7 hrs

Figure 7: Lagged correlations between cumulative rain and Lake Wanaka rise.

Knowing that the function $\mathbf{f(R_t)}$ of Eq. (6) is linear, the projection Eq. (8) was applied to the most significant two events, November 1999 and April 2010, to estimate the matrix of parameters $\propto$. Thus, the multivariate model for hourly forecast of Lake Wakatipu and Lake Wanaka levels is:

$$\begin{bmatrix} \Delta L\ (Wakatipu) \\ \Delta L\ (Wanaka) \end{bmatrix} = \begin{bmatrix} 4.28 & 0.00 & 3.3 & -101 \\ 0.00 & 5.07 & 5.2 & -19.4 \end{bmatrix} \begin{bmatrix} H \\ M \\ W \\ 1 \end{bmatrix} \quad (10)$$

The determination coefficient for the fitted model is 0.985 for both Lake Wakatipu and Lake Wanaka.

3 Model testing

Validation of the fitted models was carried out by applying the model to rainfall events which were not included in its calibration. Table 1 shows the forecasted total lake rise versus the observed one for the two events, while Figures 8 and 9 show the simulation of the fitted model to forecast hourly cumulative lake rise with a lead time of 10 hours for lake Wakatipu and 7 hours for Lake Wanaka. The figures show that the model is capable of forecasting the lake levels with determination coefficients of about 0.98 for both cases. The Filliben correlation coefficient (FC) is a little bit higher than the determination coefficient. The percentage of error ranges from 1.5% to 22%.

Table 1: Observed vs Forecasted total Lake Rise (mm).

Event Date	Lake	Observed	Forecast	R^2	FC	% Error
Nov 2001	1	357	278	0.978	0.989	22.1
	2	605	561	0.977	0.988	7.3
Dec 2010	1	424	514	0.972	0.986	-21.2
	2	1210	1192	0.968	0.984	1.5

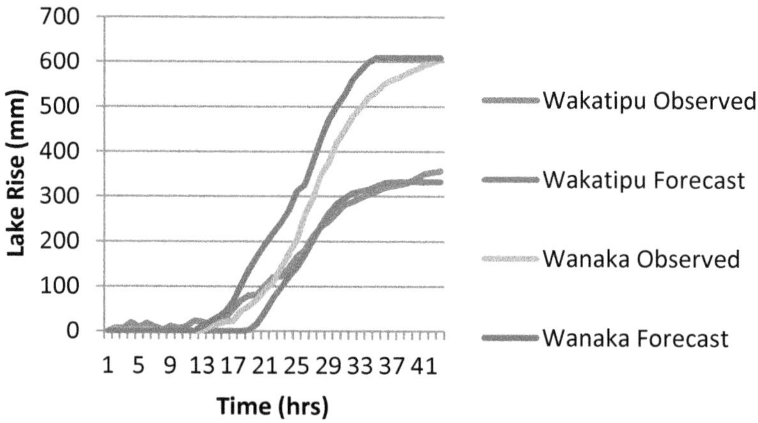

Figure 8: Observed vs forecasted cumulative lake's rise for the rainfall event November 2001.

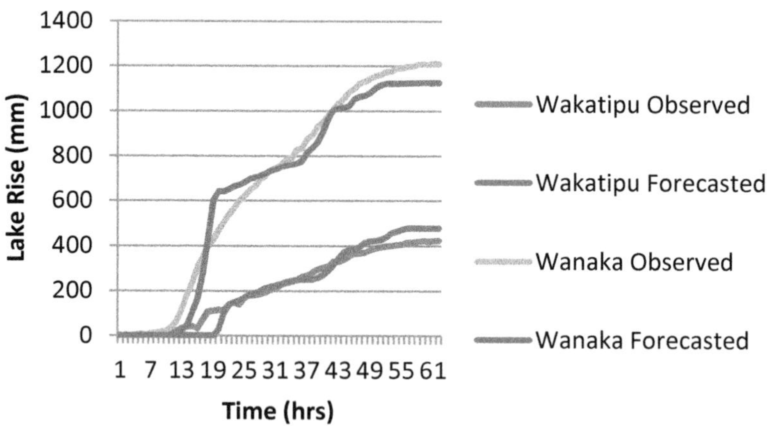

Figure 9: Observed vs. forecasted cumulative lake's rise for the rainfall event December 2010.

4 Conclusions

A multivariate model has been developed to forecast flood levels for Lake Wakatipu and Lake Wanaka in the South Island of New Zealand. The model is based on the projection theorem to obtain the optimum projection of the cumulative/total rise of lake levels onto cumulative/total rainfall at selected rainfall sites in the region. Analysis of lagged correlations between cumulative/total lake rise and cumulative/total rainfalls at the selected proved to be vital to estimate the optimum lag and to identify the function of the rainfall which is used in the projection equations. The projection equations guarantee that the produced model would produce the best forecasts. The derived model performed well in simulating observed events which were not included in the calibration process. The developed models can be used either for short term forecast during the flood event, and they give a lead time of 10 and 7 hours for Lake Wakatipu and Lake Wanaka, respectively, or they can be used for longer term forecast based on forecasted total rains at the selected sites. This modelling technique is unique and adds a new approach to the literature of flood modelling of lakes' levels.

References

[1] Bye, P. and M. Horner, Easter 1998 Floods Report by the Independent Review Team to the Board of the Environmental Agency, vol.1, Environmental Agency, Bristol, 1998.

[2] Demeritt D., H. Cloke, F. Pappenberger, J. Thielen, J. Bartholmes and Maria-Helena Ramoset, Ensemble predictions and perceptions of risk, uncertainty, and error in flood forecasting, Environmental Hazards, vol 7, pp.115-127, 2007.

[3] Jowitt, P.W. A Conceptual System Model of Rainfall-Runoff on the Haast River. Journal of Hydrology, New Zealand, 38(1), pp.121-144, 1999.

[4] Moore, R. *et al.*, Forecasting for flood warning, C.R. Geoscience vol 337, pp.203-217, 2005.

[5] Vaziri, M., Predicting Caspian Sea Surface water level by ANN and ARIMA models, JWtrwy, Port, Coast, Ocean Engineering, 123(4), 158-162, 1997.

[6] Tiwari, M. K., Chatterjee, C., Development of an accurate and reliable hourly flood forecasting model using wavelet–bootstrap–ANN (WBANN) hybrid approach, J. of Hydrology 394 (2010) 458–470.

[7] Chen-ShenHsien; Lin-YongHuang; Chang-LiChiu; Chang-FiJohn, The strategy of building a flood forecast model by neuro-fuzzy network. Journal of Hydrological Processes, 20(7), pp.1525-1540, 2006.

[8] Kisi, O., *et al.*, Forecasting daily lake levels using artificial intelligence approaches. Computers and Geosciences (2011), doi:10.1016/j.cageo.2011.08.027 [7].

[9] McSaveney, E., Floods - New Zealand's number one hazard, Te Ara - the Encyclopedia of New Zealand, updated 2-Mar-09.

[10] Mohssen, M. and Goldsmith, M., Flood Forecasting of Lake Levels: A New Concept. Int. J. of Safety and Security Eng., 1(4), pp.363–375, 2011.
[11] Brockwell, P. J. and R. A. Davis., Time Series: Theory and Methods, Springer-Verlag New York Inc., pp.46-51, 1991.

Comparison of a data-driven model and a physical model for flood forecasting

J. Ji, C. Choi, M. Yu & J. Yi
Department of Civil & Transportation Engineering, Ajou University, South Korea

Abstract

Flooding from torrential rain occurs in a short amount of time, while drought lasts for a longer period; the former may inflict huge losses in terms of both life and property. For these reasons, considerable research has been performed in the field of flood control system development. A physical model is mainly used for flood forecasting and warning. However, physical rainfall-runoff models for the conventional flood forecasting process require extensive information and data, and include uncertainties that can accumulate errors during the modeling process. On the other hand, ANFIS, which is a data-driven model combining the neural network and fuzzy techniques, can decrease the amount of physical data required for the construction of a conventional model and easily construct and evaluate a flood forecasting model using only rainfall and water level data. However, data-driven models have the disadvantage that they do not provide mathematical and physical logic, so that there are no logical correlations between the input and output data of the model. This study analyzes the characteristics of a data-driven model, ANFIS, according to its functional options and input data, such as changes in the clustering radius and the training data length. In addition, the suitability of ANFIS is evaluated through comparison with the results of HEC-HMS, which is widely used for rainfall-runoff models. In this study, the neuro-fuzzy technique is applied to the Cheongmicheon Basin using the observed precipitation and stream level data from 2008 to 2011.
Keywords: ANFIS, HEC-HMS, neuro-fuzzy technique, flood forecasting.

1 Introduction

Flooding occurs when a river overflows, usually due to torrential rain that dramatically increases runoff in a short time. Unlike drought, flooding occurs

WIT Transactions on Ecology and The Environment, Vol 159, © 2012 WIT Press
www.witpress.com, ISSN 1743-3541 (on-line)
doi:10.2495/FRIAR120111

quickly, and may cause severe damage to life and property. In Korea, precipitation is most concentrated in the summer season, and flooding frequently occurs between June and September. Steady efforts have been made to alleviate this problem, and accurate flood forecasting can be useful in this application. Some of the widely used physical models for flood forecasting include a storage function, HEC-HMS, and a unit hydrograph; however, these models have the disadvantages that the parameters require complex calculation, and many errors accumulate in the modeling process. In comparison, a data-driven model only requires rainfall and water level data from a basin for flood forecasting. Furthermore, once the model is established, it can promptly yield reliable outcomes with data input.

Advanced studies in this field are underway, including comparative analyses of flood forecasting outcomes using various methods, as well as analysis of the accuracy of the ANFIS model in relation to changing membership functions. Vernieuwe *et al.* [1] recently conducted a study to examine outcomes of a Takagi-Sugeno-type model using a changing clustering method; such an approached is a widely used rainfall-runoff model based on the data-driven technique. Chen *et al.* [2] applied the ANFIS model to forecast flooding of the Choshui River in Taiwan. They used rainfall and water level data from the precipitation observatory, and the result showed that the lasting effect and the upstream water level were essential elements in flood forecasting. Shin *et al.* [3] developed a forecasting simulator for water demand based on mathematical and neural network methods as linear and non-linear models to implement optimal water demand forecasting. It was shown that multilayer perceptron (MLP) and ANFIS, respectively, can be applied to obtain better forecasting results in multi-regional water supply systems with a large scale and local water supply systems with a small or medium scale than conventional methods.

In this study, the ANFIS model was established for flood forecasting at Wonbugyo point, which is located downstream of the Cheongmicheon Basin, a branch of the Namhan River. The precipitation and water level data between 2007 and 2011 were used as input for five rainfall events. Simulation was conducted to forecast the water level for three hours, at 10 minute intervals, starting from time T (from T+1 to T+18), and the results were compared with those of HEC-HMS.

2 Fuzzy set theory

The term *fuzzy* indicates an uncertain condition where it cannot be clearly decided whether a member of a set belongs to the set or not [4]. Fuzzy inference induces a statement from a number of fuzzy statements, and works similarly to the general inference method. For example, there are many situations that require clarification, such as "the water level of the reservoir seems *high*" and "the runoff is *substantial*." These are hard to handle with the binary codes of computer languages. Zadeh [4] developed fuzzy set theory to logically approach these ambiguous circumstances through quantitative mathematics and computer languages.

In a fuzzy set, the membership of the set has continuity. This membership can be described in three ways:

i) x belongs to fuzzy set A ($\mu_{\tilde{A}}(x) = 1$);

ii) x does not belong to fuzzy set A ($\mu_{\tilde{A}}(x) = 0$);

iii) x belongs to fuzzy set A to some degree ($\mu_{\tilde{A}}(x) = 0 \sim 1$).

ANFIS allows a user to construct a membership function with input and output data when the data are insufficient to determine the format of the membership function. Jang [5] developed ANFIS by combining the benefits of the Sugeno-type fuzzy inference system (FIS) and neural networks. ANFIS is particularly suitable for the analysis of advanced, nonlinear systems such as the rainfall-runoff model.

3 Test basin application

Rainfall and water level data from the Cheongmicheon Basin were used to compare the suitability of a data-driven model, ANFIS, and a broadly used physical model, HEC-HMS. The Cheongmicheon Basin covers the central area of the Han River system, ranging in longitude from 127°20′ to 127°44′E and in latitude from 36°56′ to 37°13′N. The basin has a triangular shape with a mild slope. The total basin area is 595.70 km^2, and the entire length is 60.8 km. The maximum east-west span is 39 km, and maximum north-south span is 31 km (Figure 1). Annual precipitation amounts to 1,061.9 mm on average, mostly concentrated between June and September. To calculate average rainfall, six

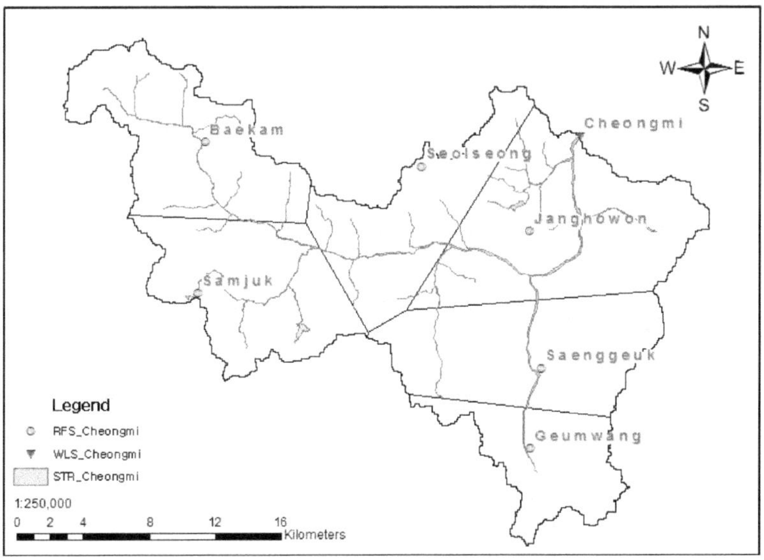

Figure 1: Map of Tancheon basin.

observatories – Saenggeuk, Seolseong, Samjuk, Janghowon, Geumwang, and Baekam – were selected, and their impact was calculated based on the Thiessen weighting method (Figure 1, Table 1). Water level data were drawn from the Cheongmi observatory located in the Wonbugyo area.

Table 1: Rainfall station of the Cheongmi basin.

Observatory	Coordinates		Thiessen Coefficient
	Longitude	Latitude	
Baekam	127° 22′ 48″	37° 9′ 36″	0.174
Samjuk	127° 22′ 28″	37° 4′ 36″	0.164
Seolseong	127° 31′ 39″	37° 8′ 45″	0.140
Janghowon	127° 36′	37° 6′ 36″	0.236
Saenggeuk	127° 36′ 25″	37° 2′ 4″	0.186
Geumwang	127° 36′	36° 59′ 24″	0.100

3.1 Model development

To construct a flood forecasting model, five datasets were selected for rainfall events that yielded substantial runoff at the Cheongmicheon Basin between 2007 and 2011. In order to examine changes due to training data length, Data 1 was further divided into Data 1_1, Data 1_2, and Data 1_3 for different lengths of input data (Table 2).

Table 2: Characteristics of selected data.

	Data Length (day)	Max Water Level (m)	Max Discharge (m^3/sec)
Data 1_1	15	4.50	588.98
Data 1_2	11	4.50	588.98
Data 1_3	5	4.50	588.98
Data 2	5	4.76	656.93
Data 3	4	3.36	326.97
Data 4	1	4.55	601.82
Data 5	2	4.74	651.60

Ten minute interval data were used for both rainfall and water level. To examine the influence of data construction, five model compositions were established for precipitation data at lead time between t and t-3, and water level data at lead times between t and t-2 (Table 3). Discharge error and RMSE were examined to compare the results (Equations (1) and (2)).

Table 3: Model composition.

	Observed Precipitation	Observed Discharge	Estimated Discharge
Model A	P_t, P_{t-1}	D_t	$D_{t+1} \sim D_{t+18}$
Model B	P_t, P_{t-1}	D_t, D_{t-1}	$D_{t+1} \sim D_{t+18}$
Model C	P_t, P_{t-1}, P_{t-2}	D_t, D_{t-1}	$D_{t+1} \sim D_{t+18}$
Model D	P_t, P_{t-1}, P_{t-2}	D_t, D_{t-1}, D_{t-2}	$D_{t+1} \sim D_{t+18}$
Model E	$P_t, P_{t-1}, P_{t-2}, P_{t-3}$	D_t, D_{t-1}, D_{t-2}	$D_{t+1} \sim D_{t+18}$

$$DE = \frac{|DP_{obs.} - DP_{est.}|}{DP_{obs.}} \times 100(\%) \quad (1)$$

DE indicates discharge error ratio (%); $DP_{obs.}$, observed peak flow; $DP_{est.}$, estimated peak flow.

$$RMSE = \sqrt{\frac{\Sigma(D_{est.} - D_{obs.})^2}{n-1}} \quad (2)$$

$D_{est.}$ represents estimated water level data; $D_{obs.}$, observed water level data; n, data length.

3.2 Changes due to different clustering radii

Clustering is a basic element in many system modeling algorithms and classifications. Its purpose is to produce compact and accurate results by grouping an extensive amount of input data by certain criteria. In the ANFIS model, a user can determine the clustering radius; the smaller the radius, the more finely the input data are grouped, yielding a greater number of rules. In this study, the clustering radius varied between 0.2 and 0.6 at intervals of 0.1; the aim of selecting these radii was to examine changes in discharge error and RMSE.

The results show that there is no significant difference in RMSE; as for discharge error, its average value was lowest (4.23%) at a clustering radius of 0.5, and the deviation from the median was also least using this parameter. The deviation was largest at a clustering radius of 0.2; as the radius increased, the deviation also gradually decreased. When the clustering radius exceeded 0.6, an error occurred, so that the system failed to produce the rules that are required to construct the ANFIS model.

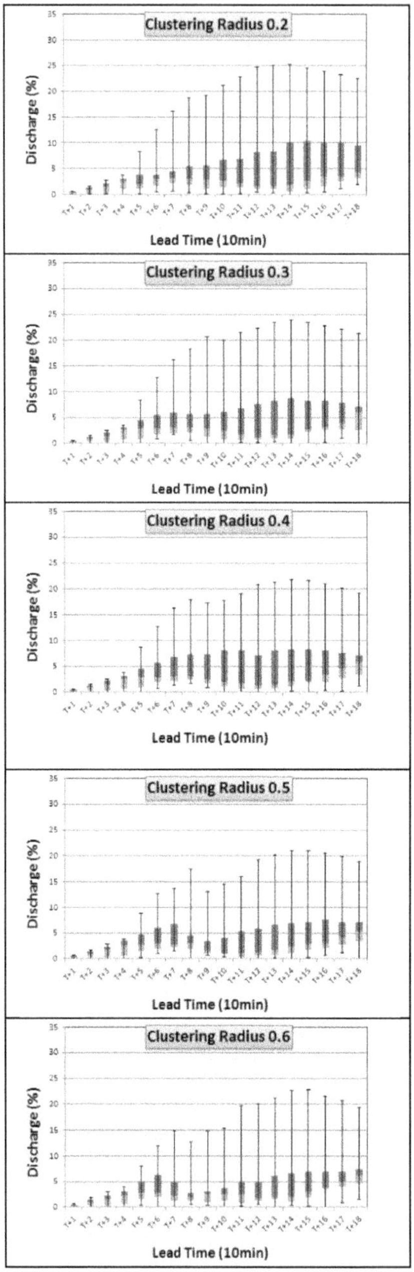

Figure 2: Discharge error comparison.

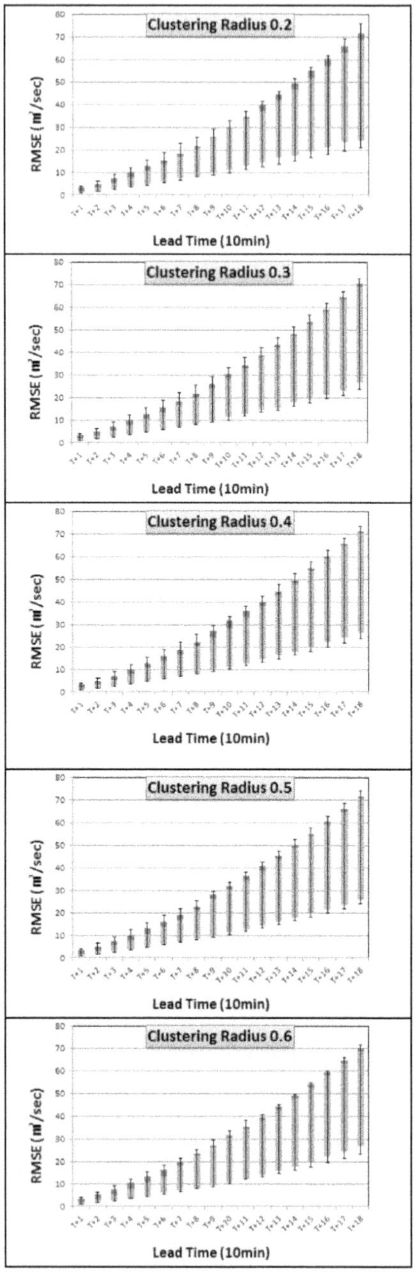

Figure 3: RMSE comparison.

Table 4: Clustering radius comparison.

	Discharge Error Average (%)	Discharge Error Median (%)
Clustering Radius 0.2	4.820	2.232
Clustering Radius 0.3	4.687	2.199
Clustering Radius 0.4	4.614	2.173
Clustering Radius 0.5	4.229	2.190
Clustering Radius 0.6	4.431	2.202

The experimental results showed that in the ANFIS model, reducing the clustering radius of the training data and producing many rules did not necessarily lead to better results in flood forecasting. It will be desirable to calculate the optimal radius for the model based on the particular characteristics of precipitation and the basin area.

3.3 Changes due to different training data lengths

As a data-driven model, ANFIS depends on training data for accuracy. It is generally known that longer training data builds a superior model. However, the observation data show that the water level may rise even when there is no precipitation due to the travel time of rainfall. To examine whether such an event adversely affects the ANFIS model, the accuracy of the model was evaluated for different lengths of training data.

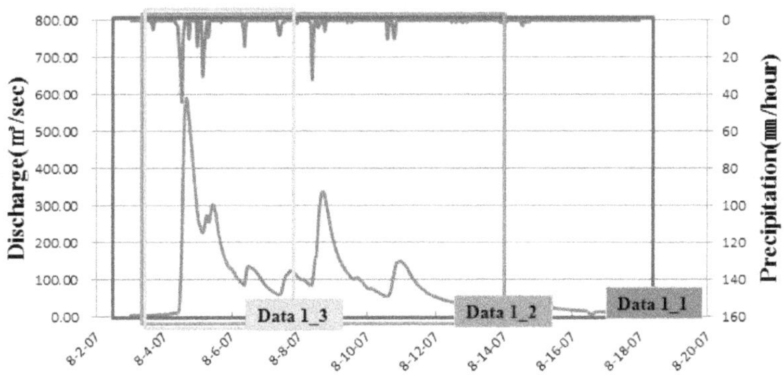

Figure 4: Training data.

Data 1_1 reflected all data for 15 days, while Data 1_2 and Data 1_3 reflected smaller amounts of data, including rainfall that showed peak flow.

The results did not show significant variance in RMSE. As for the discharge error, Data 1_3 showed a more accurate median value and a smaller range of deviation. Presumably, this is because if the water level changes when there is no precipitation, it causes noise that interrupt the rule making of the ANFIS model.

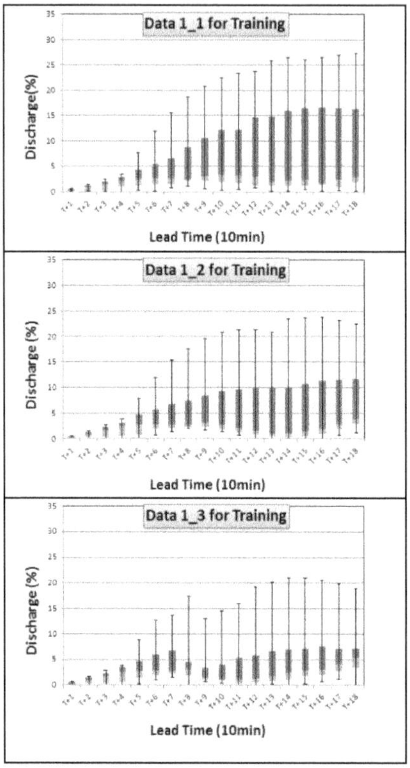

Figure 5: Discharge error comparison.

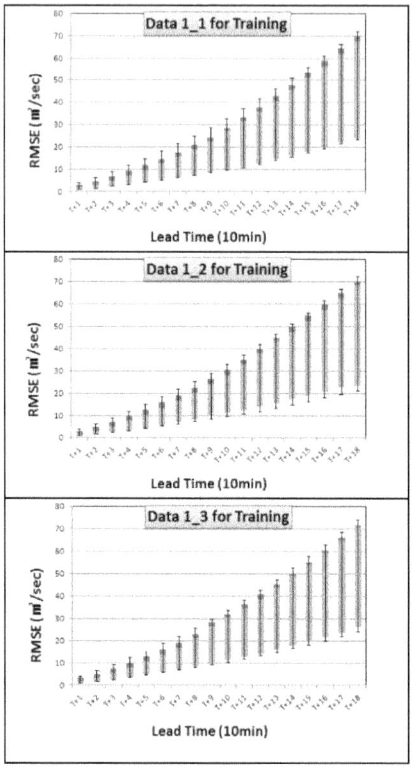

Figure 6: RMSE comparison.

3.4 Comparison of the data-driven model and physical model

Compared to the physical model, the data-driven model is relatively easy to construct and produces excellent estimates; on the other hand, it has the disadvantage of lacking mathematical and physical logic. To evaluate the suitability of the ANFIS model, its results were compared with those of HEC-HMS, a widely used physical model.

Unlike HEC-HMS, the ANFIS model makes rules with training data, constructs a model, and provides forecasts based on the observed rainfall and water level data. In this study, three rainfall events during 2011 were selected, and their observation data were compared with the forecasted outcomes of the ANFIS model and HEC-HMS. The first event took place between June 22 and 25, 2011, and was a lengthy downpour. The second event occurred on July 3, 2011 – its data were only partially taken. The last event was also concentrated in time, taking place between August 16 and 17, 2011. Figure 7 illustrates the observed discharge data and the HEC-HMS and ANFIS models' forecasts for the last event.

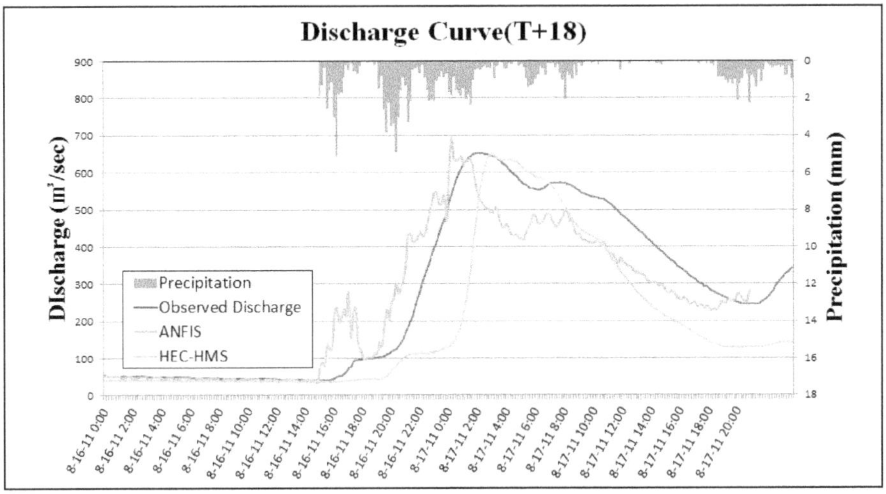

Figure 7: Discharge curve (T+18).

As the figure shows, the peak flow estimate of ANFIS at T+18 deviated further from the observation data than the discharge curve of HEC-HMS. However, overall, the ANFIS model provided more accurate estimates, as these were closer to the observed data. Table 5 shows the RMSE estimates and discharge error of the HEC-HMS and ANFIS models at T+9 and T+18.

Table 5: Comparison of ANFIS and HEC-HMS.

	June 22-25		July 3		Aug. 16–17	
	Discharge Error (%)	RMSE (m^3/sec)	Discharge Error (%)	RMSE (m^3/sec)	Discharge Error (%)	RMSE (m^3/sec)
HEC-HMS	5.25	28.25	7.25	37.39	0.40	90.15
ANFIS T+9	0.52	9.29	1.20	26.83	2.87	26.36
ANFIS T+18	4.38	23.28	6.25	70.82	6.35	71.30

As the table shows, the discharge estimate of the ANFIS model at T+9 showed excellent outcomes in most indicators, and at T+18, the estimate was similar to that of HEC-HMS.

4 Conclusion

In this study, the neuro-fuzzy technique was applied to forecast the water level at Wonbugyo point, downstream of the Cheongmicheon Basin, based on observed rainfall and water level data. Five rainfall events between 2007 and 2011 were selected to derive input data, and a model simulation was conducted for three hours (from T+1 to T+18) at 10 minute intervals. The changes in forecasting due to different clustering radii and training data lengths were examined and analyzed. The best model was compared with a widely used physical model, HEC-HMS, to identify the potentials and limitations of the data-driven model.

Changes in clustering radius did not lead to a significant difference; the result was best at a clustering radius of 0.5 in terms of discharge error. This is contrary to a general belief that a smaller clustering radius is more advantageous, as it produces more rules and thus provides a more precise forecast. The experiment suggests that the optimal clustering radius might depend on particular circumstances.

In terms of training data length, the outcome was best for the dataset that included the shortest period of non-precipitation. Generally, it is known that training data from a longer period help to build a superior forecast model. However, the experimental results suggest that water level may either rise or fall even when there is no precipitation, and such data might adversely affect the accuracy of the forecast.

In comparison with HEC-HMS, the ANFIS model produced equally excellent outcomes in terms of discharge error and RMSE; at T+9, the estimates were generally better than those of HEC-HMS. Given these results, it can be considered that the ANFIS model—as a supplementary tool to the physical model—could help to reduce uncertainties and address problems of the current flood forecast and warning system, as well as improving the accuracy of discharge forecasts.

Acknowledgements

This work was supported by a National Research Foundation of Korea (NRF) grant funded by the Korean government (MEST) (No. 2011-0029851).

References

[1] Vernieuwe, H., Georgieva, O., De Baets, B., Pauwels, V.R., Verhoest, N.E. and De Troch, F.P., Comparison of data-driven Takagi-Sugeno models of rainfall-discharge dynamics. *Journal of Hydrology*, **302(1-4)**, pp.173-186, 2005.

[2] Chen, S.H., Lin, Y.H., Chang, L.C. and Chang, F.J., The strategy of building a flood forecast model by neuro-fuzzy network. *Hydrological Processes*, 2**0(7)**, pp.1525-1540, 2006.

[3] Shin, G.W., Kim, J.H., Yang, J.L. and Hong, S.T., Development of water demand forecasting simulator and performance evaluation. *Journal of the Korean Society of Water and Wastewater*, 25(4), pp.581-589, 2011.

[4] Zadeh, L.A., *Fuzzy Sets. Information and Control,* pp.338-353, 1965.

[5] Jang, J.S., ANFIS: adaptive-networkbased fuzzy inference systems, *IEEE Transactions on Systems, Man, and Cybernetics*, 23(3), pp.665-685.

Inflow forecasting for lakes using Artificial Neural Networks

R. K. Suryawanshi[1], S. S. Gedam[1] & R. N. Sankhua[2]
[1]*CSRE, IIT Bombay, Mumbai, India*
[2]*National Water Academy, Pune, India*

Abstract

Artificial lakes or reservoirs are constructed for storing the flood waters available during the rainfall period which can be subsequently used for meeting various water demands. These storage areas are also useful for mitigating the damage caused by floods and droughts. Generally, considering the geographical/topographical aspects, many of the artificial lakes are ideally located in the upper reaches of the catchments where the rainfall is comparatively high and the submergence is comparatively low. In such small catchments, severe storms result in very high inflows entering into the storage areas in a short duration. The conventional tools for inflow forecasting include the stochastic approach in which the results depend on the quantity and quality of the input data. To overcome such limitations, the use of present computing capabilities, Satellite Remote Sensing (SRS) data, Geographical Information System (GIS) as well as Artificial Neural Networks (ANN) tools for the development of a real time inflow forecasting model based on the observed rainfall pattern to advance the forecast time as well as to achieve better forecast accuracy would help immensely in planning the lake operations.This study includes the methodology developed for inflow forecasting using geospatial tools comprising spatial and temporal analysis of the storm events i.e. digital rainfall modeling, estimation of runoff depths using ANN and conversion of runoff depths into inflows. However, this paper discusses a digital rainfall model for isochrone areas of the catchment for spatial rainfall distribution and estimation of runoff depths using ANN. An application example for the Panshet reservoir in the Khadakwasla complex near Pune (India) is used for model calibration and presented.

Keywords: digital rainfall, isochrones, ANN, GIS, SRS, inflow forecasting.

WIT Transactions on Ecology and The Environment, Vol 159, © 2012 WIT Press
www.witpress.com, ISSN 1743-3541 (on-line)
doi:10.2495/FRIAR120121

1 Introduction

Natural as well as artificial lakes or reservoirs are intended for storing the flood waters available during the rainfall period which can be subsequently used for meeting various water demands. Generally, because of geographical/ topographical constraints, the water storage structures are located in the upper reaches of the catchments where the rainfall is comparatively high. As the catchments are small, severe storms occurring in such catchments result in very high flows entering into the reservoirs in very short time duration. If the future trend in the rainfall/runoff situation is not known, reservoir water management becomes critical in view of the fact that the instant decision of storing or releasing the waters may result in an adverse impact either on the performance of the project or for the safety of the structure. To achieve proper management of the instant flood waters, it is essential to know the future trend of inflow into the reservoir. If the reservoir is operated without any inputs about the future inflow situation, it is likely that the incoming peak flood may endanger the hydrological safety of the structure as well as the life and property of the downstream population including the possibility of negation of the planned benefits of the project. Hence, there is an inherent need for every water storage/diversion project to have scientifically developed a real time inflow forecasting system on the basis of site specific data relating to all influencing parameters using the latest technological tools.

2 Reservoir inflow forecasting

The World Meteorological Organization in its international glossary of terms in hydrology (WMO [1]) defined inflow/flood forecasting as "the prediction of stage, discharge, time of occurrence and duration of inflow – especially at peak discharge at a specified point on a stream – resulting from precipitation and for snow melt". The inflow into the reservoir is dependent on meteorological, hydrological, physical and geographical parameters existing in the catchment. All these parameters have spatial and temporal variations. Hence, even if the precise and accurate forecasting is not possible, the objective is to achieve more accurate forecast using the latest tools when compared to the conventional methods being in vogue. The latest available tools include hydrologic models, GIS, ANN, etc. For such study, HEC-I, AGricultural Non-Point Source Pollution model (AGNPS) etc. are available. These models can be used to simulate hydrologic processes during flood events. However, in this study for conversion of point rainfall into spatially varied digital rainfall, spatial analyst tool of ARCGIS [2] is used. For an estimation of runoff for smaller temporal durations, ANNs have been used. Overall model calibration, testing and validation are carried out using the observed set of three hourly concurrent rainfall and runoff for the three peak storm events which occurred during the monsoon period in the study area.

3 Study area

The Panshet reservoir catchment is in the Ambi watershed of the Mutha sub basin of the Krishna river basin in Pune district of Maharashtra State in the western India. It has latitudes between 18°17'48"N and 18°23'03"N and longitudes 73°25'55"E and 73°38'00"E. The areal extent of the watershed is about 116 Sq. Km. (Figure 1) The intercepted Panshet reservoir has a storage capacity of about 300Mm^3. The time of concentration [T_c] for this elongated catchment is about 3hrs. It is observed that the flow conditions change when the floodwater enters into a reservoir as the travel time of a wave in reservoir is less than that of flood of same magnitude traveling through a natural stream. Hence, time to peak is appreciably advanced, leaving lesser warning/operation time. As there is no planned flood storage, the scope of flood moderation has been considerably limited. Hence the real time inflow forecasting model based on rainfall data is essential for scientific operation of the reservoir for safety as well as optimum use of stored water. Accordingly, the objective of the study is to develop inflow forecasting model based on the real time rainfall data using modern tools for this reservoir.

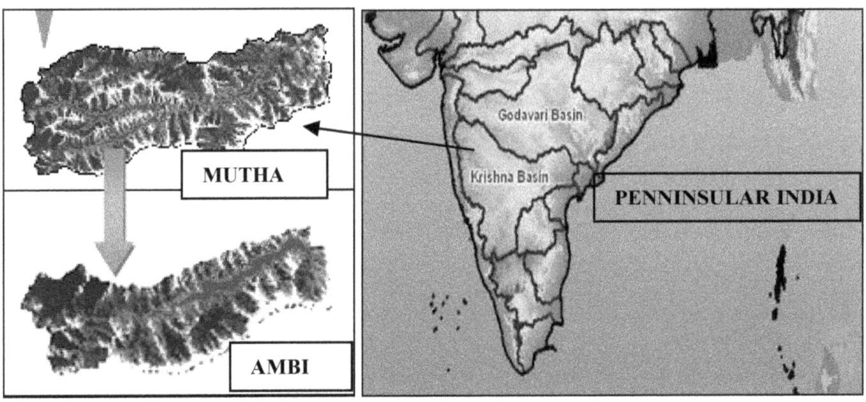

Figure 1: Location map of the study basin.

4 Methodology

The study on inflow forecasting basically involves hydrological analysis of various components like rainfall and runoff. Broadly, some of the components involved are rainfall analysis, runoff analysis, establishment of rainfall-runoff relations, establishment of catchment response function for temporal distribution of runoff and routing of flows through channels/reservoirs and establishment of operational rule curve for operation of the reservoir. The methodology flowchart is depicted at Figure 2. It is to state that the spatial distribution of the rainfall on micro scale to estimate digital rainfall, use of ANN for deriving runoff depths, and use of micro level physical and morphological features of the catchment to convert runoff depths into reservoir inflows are some of the unique features of this study. The conversion of runoff into inflows including hydrological routing duly taking into account the effect of the reservoir in computation of the inflow hydrograph at the dam site using geospatial tools, and calibration, testing and validation of the model for the study basin are part of this flowchart. However, in this paper the discussion is limited to estimation of digital rainfall using innovative method using advance geospatial tools and estimation of runoff using ANN including its validation using observed events for the study area. The remaining components are under study.

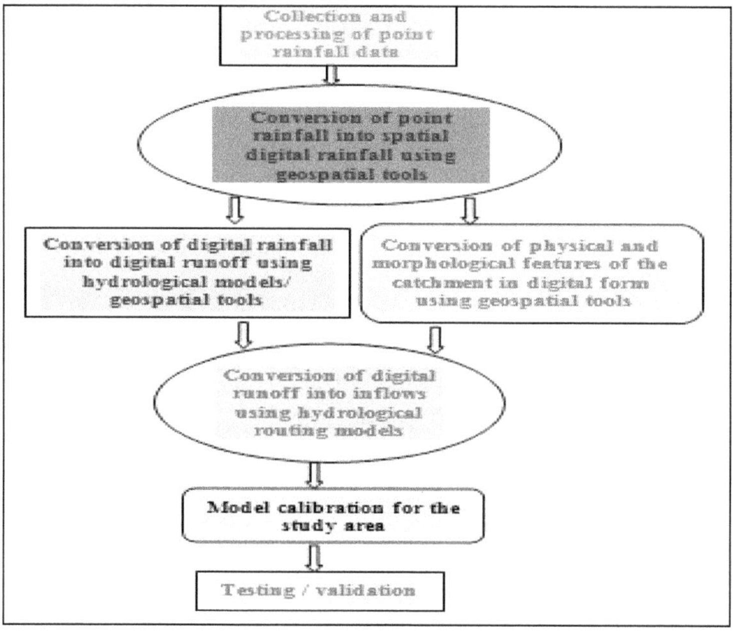

Figure 2: Methodology flowchart.

5 Digital rainfall estimation

The point rainfall being the fundamental input in the process, the accuracy in estimation of the average areal rainfall that falls in the catchment during the storm is the primary concern for forecasting the inflows into the reservoir. Ball and Luk [3] have attempted to model the rainfall variation, whereas Narkhedkar *et al.* [4] used spatial grid for the same. The mean areal value (R) of rainfall over a basin area *A*, having pixel element ($dx_, d_y$) can be expressed as

$$\boldsymbol{R = \int_A r(x, y).d_x.d_y / \int_A d_x\, d_y} \tag{1}$$

where $\boldsymbol{r(x, y)}$ is the rainfall surface function which is never known but can be approximated from the point rainfall values $r_{1,}$ $r_{2,}$,…, r_N measured by gauges.

The simplified results for average areal rainfall can be obtained using geospatial tools. The point rainfall of the four stations collected for the study area is distributed spatially using the innovative concept of Digital Rainfall Model (DRM) which is similar to the concept of Digital Elevation Model (DEM). The pixel wise spatial distribution of rainfall is estimated for each of the three hourly spells of the storm using Inverse Distance Weights (IDW) available in spatial analyst tool in ARC-GIS. The distributed rainfall is again classified in discrete intervals for each of the isochrone polygons which are subsequently converted into average rainfall for the entire catchment based on the all isochrone polygon areas. The sample DRM layer is shown in Figure 3 and the three hourly distributed rainfall hyetograph for the storm event is depicted at Figure 4.

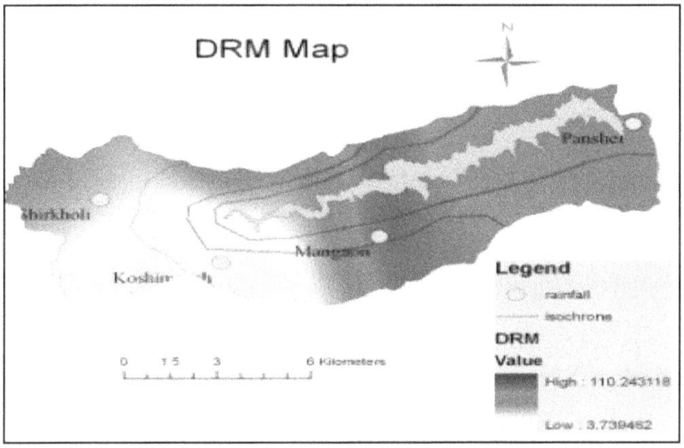

Figure 3: DRM and isochrone layer.

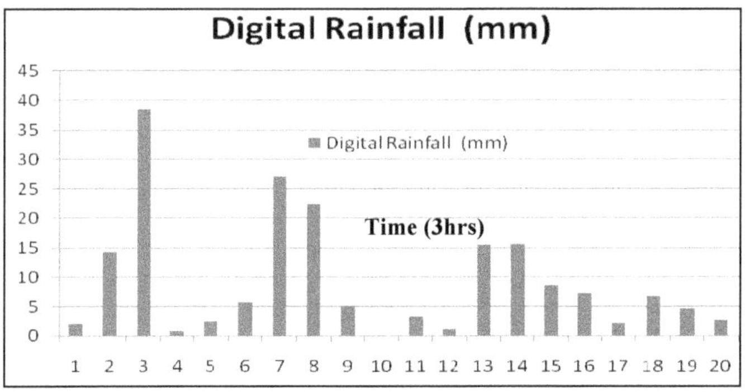

Figure 4: Digital rainfall hyetograph.

6 Runoff estimation using ANN

The other important step in inflow forecasting is estimation of runoff depths using appropriate hydrological models. The use of ANN in hydrological analysis has become quite common. The technique is being efficiently used as black box model for estimation of various components of storages and processes involved in hydrologic cycle. Tokar and Johnson [5] have attempted to use an ANN methodology to forecast daily runoff as a function of daily precipitation, temperature and snowmelt. Sajikumar and Thandaveswara [6] used artificial neural network paradigm i.e. temporal back propogation-neural network (TBP-NN) for estimation of runoff from rainfall data on monthly basis. Zealand *et al* [7] have used ANN for short-term forecasting of stream flows. The model of an artificial neuron closely matches biological neuron. The artificial neuron is a processing element, i.e. a node. The input signals $x_1, x_2, x_3 \ldots x_n$ are normally continuous variables. The weights can be positive or negative corresponding to acceleration or inhibition respectively of the flow of the signals. The summing node gathers all the input weighted signals and then passes to the output using the transfer function which is usually non-linear. Applying the most commonly used sigmoidal transfer function, the output of typical neuron can be written as

$$y = 1/(1+ w^{-ax}) \tag{2}$$

where $\boldsymbol{x}$ is the input, $\boldsymbol{y}$ is the output, $\boldsymbol{w}$ is the weight and $\boldsymbol{a}$ is the coefficient/gain which reflects the slope of the sigmoidal function that changes between the two asymptotic values (0 and +1). Training is a process of changing/updating the weights. The back propagation training algorithm is used for a wide variety of applications.

7 Design of ANN network

NeuroIntelligence (neural network software) [8] is used in this study to solve the forecasting problem which helps to gain the maximum productivity in preprocessing data, efficient network architecture, analyze performance. Network with five nodes in input layer, eight nodes in one hidden layer and one node as output has been adopted for the study (Figure 5).

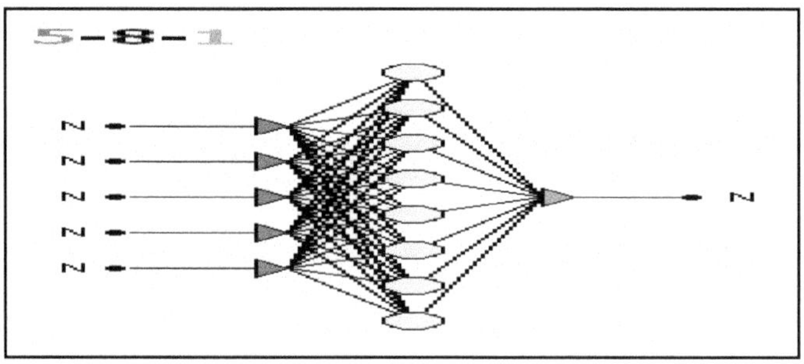

Figure 5: The ANN designed network adopted.

The point rainfall data of each of the four stations as well as the digital average rainfall obtained as above is used as an input data for the ANN network (Figure 6). The corresponding observed three hourly runoff depths have been used for training, testing and validation of the ANN network.

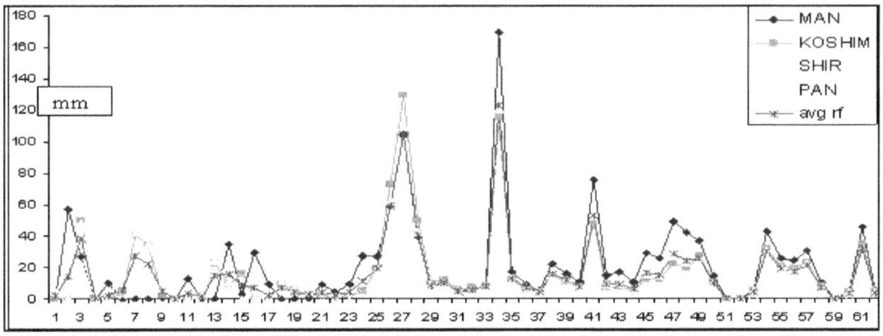

Figure 6: Sample input rainfalls.

8 ANN output results

The training, testing and validation results are compared with the concurrent observed runoff depths computed using the concurrent reservoir levels. The results obtained are depicted in Figure 7.

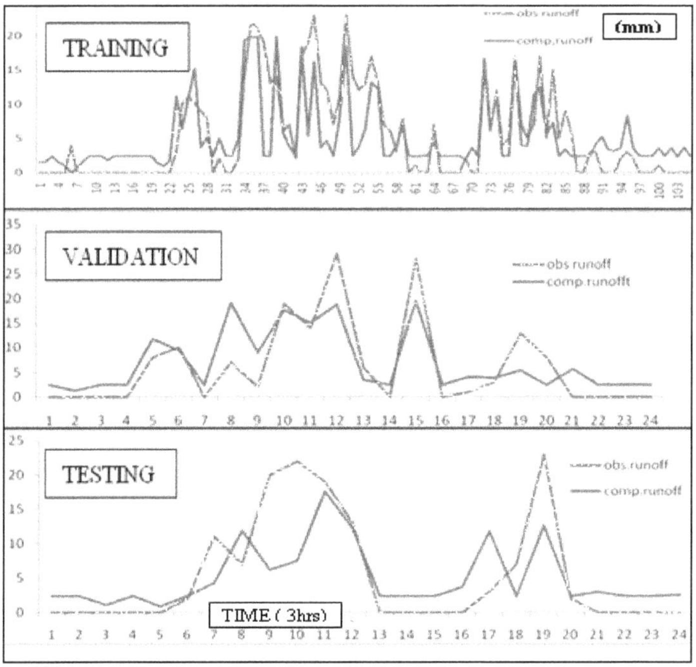

Figure 7: ANN training, validation and testing results.

The sample query average rainfall hyetograph and the corresponding predicted runoff is shown in Figure 8.

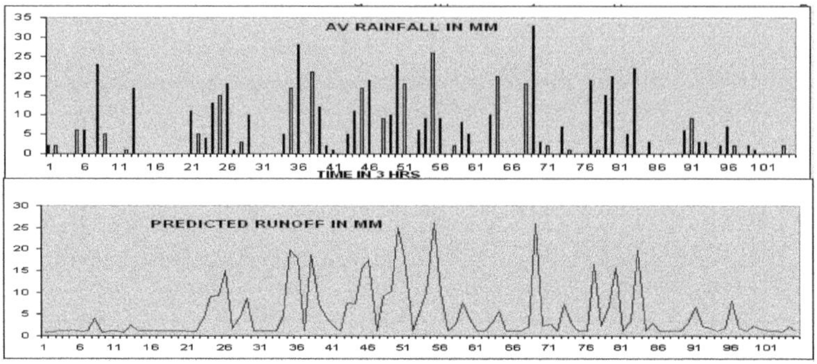

Figure 8: Sample query results.

9 Conclusion

The output results as shown in Figure 7 indicate that the observed and computed runoff depths are more or less matching with respect to peaks as well as time. Hence it can be inferred that the design as well as the training, validation and testing of the ANN network is quite satisfactory. The results of the sample query input hyetograph as shown in Figure 8 are also satisfactory. Thus it can be inferred that the trained ANN network with the observed rainfall depths as an inputs can be very well used for runoff forecasting with more accuracy and very short processing time. Further research on remaining aspects of the reservoir inflow forecasting methodology flowchart is under progress.

References

[1] WMO-UNESCO-XXIV *"International Glossary of Terms in Hydrology"*-WMO (1974)/ (385-II edition- +413PP).

[2] Schumann A.H., R. Funke, G.A. Schultz *"Application of GIS for conceptual rainfall runoff modelling"* Journal of Hydrology 240 45-61(2000).

[3] James E. Ball and Kin Chai Luk *"Modeling Spatial Variability of rainfall over a catchment"* Journal of Hydrologic Engineering ASCE, 122-130. April 1998.

[4] Narkhedkar *et al.* *"Rainfall analysis using conventional and non-conventional rainfall information on monthly scale"* Atmósfera 23(2), 141-164(2010).

[5] Sezin Tokar, Peggy A. Johnson *"Rainfall – Runoff Modeling using Artificial Neural Network"*, – ASCE Journal of Hydrologic Engineering Vol4 No.3 Paper No. 14723 (232-239) July 1999.

[6] Sajikumar N., B.S. Thandaveswara *"A Non-Linear rainfall – runoff model using artificial neural network"* Journal of Hydrology 216 32-35(1999).

[7] Cameron M. Zealand, Donald H. Burn, Slobodan, P. Simonovie *"Short Term Stream flow forecasting using artificial neural networks"* Journal of Hydrology 214 32-48 (1999).
[8] http:/www.alyuda.com/neural-networks-software.htm

Section 5
Flood case studies

Impacts and preventative measures against flooding and coastal erosion in Thailand

S. Pongpiachan[1], T. Settacharnwit[1], P. Chalangsut[1],
P. Hirunyatrakul[2] & I. Kittikoon[2]
[1]NIDA Centre for Research and Development on Disaster Prevention and Management, School of Social and Environmental Development, National Institute of Development Administration (NIDA), Thailand
[2]Bara Scientific Co., Ltd., Bangkok, Thailand

Abstract

The purpose of this paper is to describe the natural disasters of flooding and coastal erosion in Thailand, which have resulted in many deaths and large-scale property damage. The magnitude and frequency of Thailand's flooding in the past and the present are also examined.

The MIKE 11 and URSB hydrological models are shown to be the most suitable models for forecasting floods in Thailand, as well as for planning short and long term protection and mitigation measures.

With its long length of coastline on both the Gulf of Thailand and the Andaman Sea, Thailand faces the increasing occurrence of coastal erosion. The major causes of coastal erosion originate from both natural processes and human activities. A simulation of a rise in sea level by incremental degrees and their potential impact on some major cities in Southeast Asia is described and measures to mitigate and limit the negative impact of coastal erosion are presented in order to preserve the environment for future generations.
Keywords: flooding, coastal erosion, Thailand, hydrological model, sea level rise, mitigate and adaptation limit.

1 Introduction

The threat of environmental problems has become an increasingly more pressing issue in Thailand, owing to the tremendous negative impact on human life and property. Thailand is subject to many natural disasters including floods, tropical

WIT Transactions on Ecology and The Environment, Vol 159, © 2012 WIT Press
www.witpress.com, ISSN 1743-3541 (on-line)
doi:10.2495/FRIAR120131

cyclones, earthquakes, thunderstorms, landslides, storm surges, coastal erosion, sea level rises, fires and drought (TMD [1]). Meanwhile, human activities also deplete natural resources at an ever-growing rate. Deforestation and industrialization have led to changes in land use from agricultural to residential and commercial, which has also had a negative impact on the biodiversity in the region as pointed out by McDonald [2] and Chunkaew [3]. In many situations, the eco system has the ability to rehabilitate itself or show its resilience by bouncing back after an external disturbance that is not too drastic (Miller [4]). However, its vulnerability is exacerbated when combined with the destructive impact of human activities (IPCC [5]).

This paper aims to present the disasters of flooding and coastal erosion in Thailand, which have occurred more frequently and violently over the last decade. The flooding at the end of 2010 overwhelmed 51 provinces, leaving 258 dead and 8.97 million people affected (Nirapai Center [6] and CRED [7]). Likewise, 11 percent of the coastline along the Gulf of Thailand and two percent of the Andaman Sea coastline are eroding at a rate of more than five meters a year (or two square kilometers of coastal real estate). The damage has been estimated at approximately $US 156 million/year (Jarungrattanapong and Manasboonphempool [8]). The causes of flooding and coastal erosion are examined and recommendations for short and long term prevention and mitigation measures are presented in this paper.

2 Flooding

Flooding is the single most destructive type of natural disaster that strikes humans and their livelihoods around the world. Agencies provide a variety of definitions of flooding, as follows: The United Nations, International Strategy for Disaster Reduction (UNISDR) explains that "flooding can arise from overflowing river, heavy rainfall over the short period or an unusual inflow of sea water onto land, such as storm surge, tidal flooding, seismic events (tsunami) or large landslide" (Pilon [9]).

The United States Environmental Protection Agency (USEPA) defines universally that "flood or flooding means a general and temporary condition of partial or complete inundation of normally dry land areas from the overflow of inland and/or tidal waters and/or the unusual and rapid accumulation or runoff of surface water from any source or flooding from any other source." Meanwhile, the European Union (EU) Directives define flooding as "the temporary covering by water of land not normally covered by water." This definition include floods from "rivers, mountain torrents, Mediterranean ephemeral water courses, and floods from the sea in coastal areas," but and may exclude floods from sewerage systems. The American Meteorological Society defines the meaning of flooding as the "overflowing of the normal confines of a stream or other body of water, or the accumulation of water over areas that are not normally submerged." These attempts illustrate the difficulty in arriving at a precise definition of "flood or flooding."

In the case of Thailand, there are three types of flooding that can be characterized by pattern and severity as follows: 1) *general or regular flooding*, characterized by the unusual presence of water on normally dry lands. 2) *flash flooding*, which occurs with heavy rainfall over a short period of time, and 3) *river flooding*, used to describe overwhelmed or overflowing rivers (TMD [10]). However, Thailand is also confronted by the more rarely-occurring *ocean flooding*, defined by a sudden inflow of sea water onto land, such as the tsunami created by the earthquake in the Indian Ocean on December 26, 2004 which caused large-scale social and economic damage in Thailand (UNESCO [11]).

2.1 Flooding and climate change

Several models have been used to investigate the impact of climate change on the magnitude of flooding in different regions. For instance, the Nicholls model predicts the average annual people flooded (AAPF) based on three different scenarios – namely low-, middle-, and high sea-level rises – as a consequence of climate change [12]. Although major uncertainties were found in the model, the relatively strong tendency towards the increase of the AAPF as a result of sea-level rises is predominant. It is also worth mentioning that the incidence of flooding triggered by sea-level rises greatly depends on the geophysical conditions of individual regions.

In addition, the occurrence of flooding in West Africa, East Africa, the Southern Mediterranean, the Southeast Asia, and the South Asia will be intensified and worsen if there are no actions on reducing green house gases (mitigation) and developing of the coastal planning and management (adaptation) (Nicholls and Mimura [13]). In addition, sea level rises and flooding of coastal lands will lead to salination or contamination of fresh water and agricultural lands, and the loss of nursery areas for fishing (Costello *et al.* [14]). Since the intensity of river overflow is a function of water run-off volume and precipitation, it is therefore crucial to assess the influence of these two parameters on occurrences of river flooding (Kozka and Holko [15]). The impact of climate change on river flooding was estimated by combining the stochastic precipitation model with current climate change conditions (Booij [16]).

There are a number of hydrological basin models developed for flood forecasting, including TOPMODEL, HBV, MIKE 11, URBS and GCMs. The criteria for model selection are comprised of objectives/purposes of flood forecasting, required data for each model, and the conditions of topography characteristics, catchments, estuaries, run-off and rainfall volume. For example, UBC, the Canadian model used for forecasting run-off from mountain catchments, has the advantage of estimating depletion and operating entirely from meteorological inputs of daily maximum and minimum temperatures and precipitation. However, its disadvantage is the difficulty of collecting scarce meteorological and sparse stream data in mountain areas, if available, a more extensive database can be utilised. If the objective is to forecast global weather patterns, GCM should be considered, since it can operate under various conditions but has large biases. HBV is applied to various different climate conditions while MIKE 11 is used for multi-purpose forecasts.

Some of the hydrological models have been applied to planning for the protection from and mitigation of flooding disasters in the Southeast Asia sub-region.

The URBS model was applied for flood forecasting in LMB instead of existing forecasting models that are out of date and unreliable. Furthermore, users can make use of URBS' simple recovering initial loss rainfall runoff model or integrate the results from a third party water model (MRC [17], Malone [18] and Carroll [19]). In Thailand, the Danish hydrodynamic model, MIKE 11 is considered the most suitable model for forecasting and management of floods since it can be applied to many purposes.

The standard MIKE 11 model provides applications for long-term forecasting accuracy that include: 1) real-time monitoring and decision support 2) real-time flood forecasting and warning system 3) control of dams, reservoirs and hydraulic structures, and 4) water resource and environment monitoring (Skotner *et al.* [20]). Nevertheless, there are several disadvantages to MIKE 11 such as numerical instabilities when river segments run dry and water balance errors (Narrabri Shire Council [21]). The Water Crisis Prevention Center has utilized MIKE 11 for flood forecasting in the Chao Phraya River Basin and has employed this model to calibrate and verify the extent of the flood disaster which occurred in 2006 and 2007. MIKE 11 was used for flood forecasting in 2008 to compare the observed results with data from the study area and statistical values. The results were quite acceptable and very reliable, and the system was used again to simulate one-dimensional flow in the Nam Loei River together with the generation of a map of the flooding areas (Chitprom and Patsinghasaneel [22] and Taesombat *et al.* [23]).

2.2 Thailand's flooding situation

Thailand is located at 17°00′N 100°00′E/17°N 100°E. Its topography is characterized by fertile floodplains with a tropical monsoon climate in the central area, marginal uplands and mountains in the Northern and the Northeastern regions, and low lands with long mountainous outcrops between the Gulf of Thailand and the Andaman Sea. In the case of Thailand, floods have occurred frequently since the Sukhothai regime (1257–1420 A.D.). Some stone inscriptions record the occurrences of floods at that time caused by the very large floodplain area and the huge volume of water flowing from the mountains in the extreme North which was unable to discharge into the Chao Phraya river.

The major flooding of Bangkok in 1942 was recorded as having effects on the city's infrastructure lasting three months; the trams had to stop operation downtown and boats had to be used for commuting to work. The major causes of this flooding were: 1) Bangkok's location on the central, flat flood plain of the Chao Phraya river basin with a low elevation above sea level, as shown in fig. 1, and 2) having heavy continuous rainfall before dams had been built in the northern region; therefore, it was difficult to restrain the high magnitude of water discharged before it reached Bangkok.

Thailand experienced severe flooding from October to December 2010 (fig. 2). The flooding was caused by the early arrival of the La Niña phenomenon

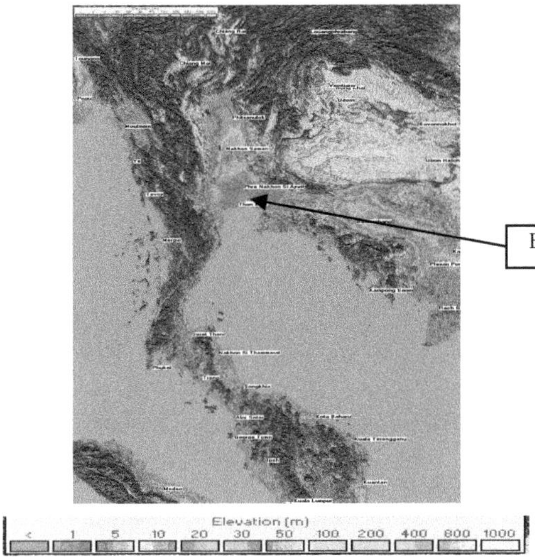

Figure 1: Map of elevation relative to sea level (m), Thailand (Source: www.globalwarmingart.com/wiki/Special:SeaLevel.)

Figure 2: Flooding in the year 2010 compared with flooding in the past.

and heavy rain in many areas. It overwhelmed more than half the country's regions (51 of 76 provinces), leaving 258 people killed, and 8.97 million people affected. More than two million households and 3.2 million acres of farmland were affected; the rapid response flood inundation map is depicted in fig. 3.

The highest frequency of flash flooding occurred in the Northern and the Southern areas, followed by general flooding and storm surge/coastal flooding in turn; however, the general flooding, flash flooding and storm surge/coastal flooding were also found, but less frequently, among the North Eastern, the Western and the Eastern regions (CRED [7]).

Since the northern part of Thailand has always experienced flash flooding, there is a wealth of appropriate local knowledge People know how to protect against floods and reduce water damage by building their houses in the

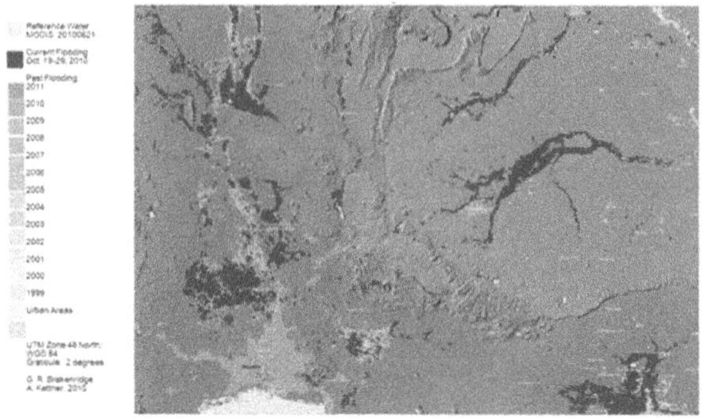

Figure 3: Rapid response flood inundation map, Thailand, October 10, 2010. (Source: www.floodobservatory.colorado.edu/hydrography/E100N20 RR1.html.)

traditional Thai style. The house has a high floor level allowing a man of average-height to walk with clearance above his head. The space under the house is used for storing farm equipment, keeping livestock, producing handicrafts, and as a common sitting or squatting area.

Due to the specific geophysical conditions of the Southern region, most areas used to be surrounded and covered by tropical rainforest. Furthermore, their tap roots and fibrous roots could capture soil at a deeper level to extend water retention time, protect against soil erosion and landslides. Nowadays, the rainforest areas have been reduced and substituted by rubber and palm oil plantations. The rubber tree tap roots have an average depth of no more than 1.5–2.0 meters (Wannasiri [24] and LDD [25]). Thus, when severe flash flooding occurred, most rubber trees were uprooted easily. Intrusion into mangrove forest areas by palm oil plantations, grown as an energy-producing crop has led to the reduction of these natural flooding buffer areas. Consequently, the combination of these events caused an extremely large amount of water from heavy rain to inundate the city and trigger landslides to lower areas.

The obvious example of this case is how the heavy rains from the tropical depression through the Gulf of Thailand brought serious flooding to Hat Yai, in Songkhla province, the main commercial district in the Southern part of Thailand in November, 2010, as shown in fig. 4.

To mitigate the catastrophic effects of flooding, all concerned parties should cooperate on setting short and long term plans and co-ordinate their action plans for flood management. In the short term, the planting of Vetiver grass in rubber plantations should be promoted, which can decelerate water velocity. The local authority should install early warning systems for floods and other natural disasters for protection and to enable people to be evacuated from risk areas in time. In the long run, to achieve more effective practical policies, the state policies on rainforest and mangrove reforestation for flood mitigation, coupled with city planning to relocate commercial zones to uphill areas to reduce property damage have to be implemented.

Figure 4: The damage of flooding in Hat Yai District, Thailand. November 2, 2010. (Source: www.reuters.com and www.bangkokpost.com.)

In the case of the city of Hat Yai, the government should consider adopting the URBS model which can be utilised to model a specific event or perform continuous simulations to help mitigate current conditions and events. Moreover, the conditions of the model are consistent with flood forecasting in the case of rare, high-rainfall events. Besides these advantages, URSB also allows users to configure the model to match the individual characteristics of catchment areas. The early warning system must be installed in parallel to reinforce the evacuation plan.

3 Coastal erosion

Coastal erosion is a major issue for countries that have coastlines, especially in most Pacific Islands; a solution to this problem needs to be found urgently or at least measures to alleviate the damage (South [26]). The UK's Department of Environment Food and Rural Affairs has defined coastal erosion as a natural process by which coastlines adjust to varying sea levels, energy levels, sediment supply and existing topography. Over most of the ten millennia of the Holocene Period (characterized by generally rising global sea levels), coastlines have undergone substantial erosion and deposition to gradually assume their current configuration. Some continue to evolve through large-scale redistribution of sediment through erosion and deposition. Management problems arise when fixed human infrastructure is threatened with destruction by coastal erosion" (Cooper and McKenna [27]).

Coastal erosion is defined briefly by the South Pacific Sea Level and Climate Monitoring Project as "the loss of land along the shoreline. It happens when sediment, in the form of sand and soil, is removed from the coast. Coastal erosion is a serious problem since it has a negative effect on the way of life, grouped settlements and economic activities of people who live near low-lying coastal land (UNEP [28]). Overall, we can conclude that the general causes of coastal erosion arise from both natural and human activities.

3.1 Coastal erosion in Thailand

Due to the long length of sea coast on the Gulf of Thailand (1,660 kilometers stretching along 17 provinces, including Bangkok) and the Andaman Sea (2,614 kilometers), Thailand is confronted by a high risk of land erosion along the coasts. The high erosion rates, at an average of five meters annually, have been measured across 180.9 kilometers in 12 provinces or 10.9 percent of the Gulf of Thailand's coastal length (Maneerat [29]) as well as across five out of six provinces along the Andaman coastline for 90.5 kilometers out of 954 kilometers or 9.5 percent (Maneerat [29], Suvanich [30]).

3.2 Areas affected by sea level rises

A hypothetical situation where an incremental rise in the sea level in Thailand was simulated at zero, five, and 10 meters is shown in fig. 5(a)–(c), taken at the Jet Propulsion Laboratory, California Institute of Technology [31] and the details can be explained below.

At a five-meter rise in sea level, the simulation showed that some provinces in the central part of Thailand (Bangkok, Nonthaburi, Samutprakarn and Pathumthani provinces) and some areas on the Gulf of Thailand (the upper part of the Southern peninsula, mainly in Prachuabkirikhan, Chumporn, Nakorn sithammarat and Songkhla provinces) would be immersed. This would have a similar effect on the Southern part of Myanmar (Rangoon), the Southern part of Cambodia and also Hochiminh city in Vietnam.

As a result of a sea level rise of 10 meters, larger areas of the central region, i.e. the whole province of Samutsakorn and Samutsongkram, and some areas of Ayuthaya and Aungthong, would be submerged. More of the Eastern part of the Gulf of Thailand would be affected by this sea level rise (i.e. the Eastern part of Petchburi, extending south to Surathani and Samui Island). Its effects would extend to the Andaman Coast as well, especially some areas on the western coast of Malaysia, Central and Southern parts of Myanmar and Cambodia. This simulation predicts that Phanom Penh city would disappear; nevertheless, this level of sea rise would have no effect on landlocked Lao PDR.

3.3 Measures to mitigate and protect against coastal erosion

Rattanamanee *et al.* [32] revealed that the major causes of coastal erosion in Thailand emerge from both natural processes and human activities. Natural processes, including numerous incidents such as rising sea levels, shrinking residential areas, changing of the balance of shoreline sediment and climate change are considered crucial issues. It is also important to note that several human activities such as extracting sand from beaches, mining coral, removing mangroves, dredging channels, and constructing structures such as seawalls, groins and causeways can tremendously alter tidal circulation and wave processes around islands. Eroded land affecting public and private properties can be documented, the total extent of the damages – which includes the negative impact on the balance of coastal ecology and biodiversity, on coastal communities' quality of life, and on the country's gross economy – is difficult to evaluate.

The subsequent recommendations are suggested for the on-going mitigation and prevention of coastal erosion: First, relevant government agencies should establish a comprehensive policy or strategy roadmap to actively promote the prevention and remediation of damaged coastlines. This plan should allow all stakeholders the opportunity to participate in the decision-making at every stage in the process. Secondly, the most appropriate technical interventions should be chosen for each of the coastal areas based on the needs and conditions. For example, as a short-term solution, it is essential to construct seawalls, breakwaters and sand sausages to hold back waves from attacking the coast, carry outbeach and dune nourishment and artificial coral reef barrier construction to reduce tidal power before it strikes the coast.

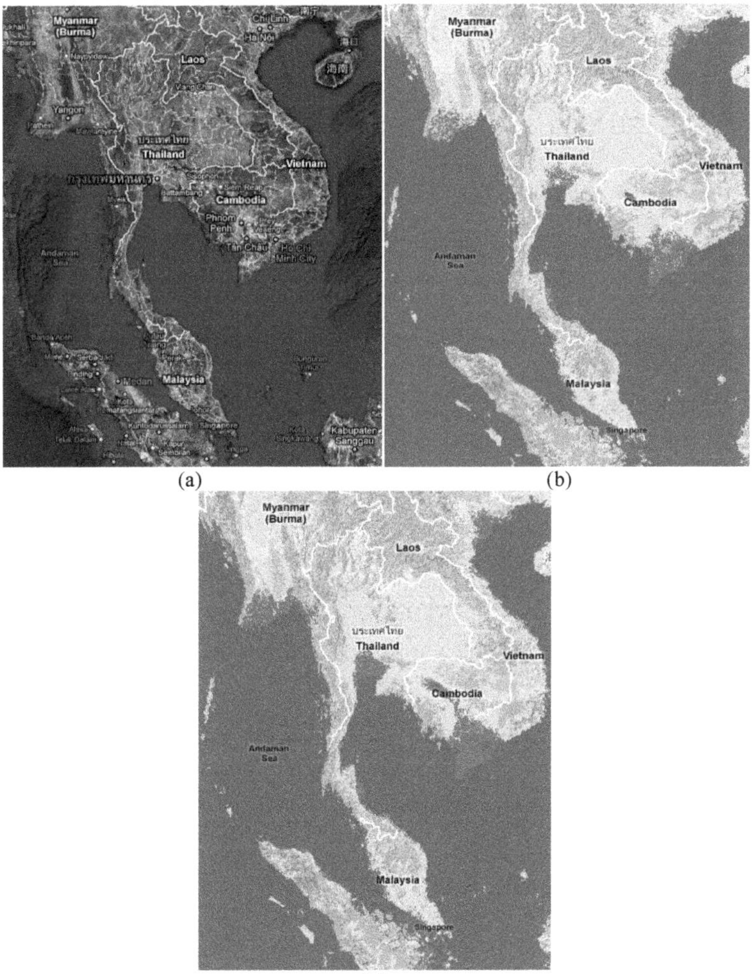

(a) (b)

(c)

Figure 5: (a) Sea level rise + 0 m. (b) Sea level rise + 5 m. (c) Sea level rise + 10 m. (Source: Jet Propulsion Laboratory, California Institute of Technology (2011). Accessed from www.flood.firetree.net/?ll= 10.5310,99.0967&z=12&m=14&t=3

In addition, results provided by Thampanya, *et al.* [33] indicate that mangroves could reduce the mean erosion rate from 3.3 $\pm$ 0.4 m y^{-1} to 2.3 $\pm$ 0.4 m y^{-1}as for the coastal segment without mangroves. As a long term solution, the natural environmental features of the area should be re-established, especially mangrove reforestation. In the case of severely eroded shorelines, Jarupongsakul [34] and Jarungrattanapong and Manasboonphempool [35] have proposed that it is necessary to launch measures by combining both short and long term solutions against the threat from sea level rises in order to preserve a sound environment for the benefit of future generations.

4 Conclusion

To mitigate against environmental disasters, the hydrological models have been applied to simulate river or stream flow and calculate water quality parameters. In Thailand, the hydrodynamic models, MIKE 11 and URSB, are considered appropriate models for flood forecasting and management.

This paper has established that flooding has been more frequent and violent in the last decade. Human activities are playing a key role in increasing the number and severity of natural disasters, especially through the reduction of rainforest areas and substitution by rubber and palm oil plantations, road construction and waterway diversion.

In addition, coastal erosion in Thailand is found on both the Gulf of Thailand and the Andaman Sea coasts. The Geographic Information System software simulation of sea level rises provided by California Institute of Technology shows that some areas of Thailand, especially in the Central and Southern regions would be immersed in the event of higher sea levels of more than five meters occurring. In the case of sea levels rising by 10 meters, the cities of Bangkok, Rangoon, Hochiminh and Phnom Penh would vanish from the map.

To effectively deal with the natural disasters of flooding and coastal erosion, all interested parties from private, government and non-government sectors should cooperate on setting short and long term plans, and their action plans by adjusting them to appropriately serve specific areas.

Proactive measures (city plan zoning, mangrove reforestation) and reactive measures (warning system installation and evacuation plans) should be set up to reduce the impact and minimize the loss of life and damage to property.

In conclusion, natural disasters will still occur and are difficult to avoid. However, resolutions of these disasters are to have good preparation and management including avoiding over-exploiting the natural resources.

References

[1] Natural Disasters in Thailand, Thai Meteorological Department (TMD), www.tmd.go.th/info/info.php?FileID=70

[2] McDonald, R., Introduction to Natural and Man-made Disasters and their Effects on Buildings, Architectural Press, Oxford, UK, pp.1-17, 2003.

[3] Chankaew, K., *Environmental Science,* (8th eds), Kasetsart University Press, Bangkok, pp.22-23, 2008 (in Thai).
[4] Miller, G.T. Jr., Community ecology: structure, species interactions, succession, and sustainability (chapter 7). *Environmental Science,* 10th edn., Thomson Learning, Inc., pp.140-159, 2004.
[5] Bates, B.C., Kundzewicz, Z.W., Wu S. & Palutikof, J.P., Executive summary. *Climate Change and Water*, Technical Paper of the Intergovernmental Panel on Climate Change, IPCC Secretariat, Geneva, pp.1-4, 2008.
[6] Nirapai Center, *Flooding Situation, 2010.* www.disaster.go.th/dpm/flood53/flood53 .html
[7] The International Disaster Database, Centre for Research on the Epidemiology of Disasters (CRED), www.emdat.be/result-country-profile
[8] Jarungrattanapong, R. & Manasboonphempool, A., Adaptation strategies for coastal erosion/flooding: a case studies of two communities in Bang Khun Thian District, Bangkok., *TDRI Quarterly Review*, pp.11-18, March, 2008.
[9] Pilon, P.J. (eds), *Guidelines for Reducing Flooded Losses,* United Nations, International Strategy for Disaster Reduction (UNISDR), pp.3-4, 2002.
[10] Thai Meteorological Department-TMD, Natural Disasters in Thailand, www. tmd.go.th/info/info.php?FileID=70
[11] United Nations Educational, Scientific and Cultural Organization- UNESCO, Case studies: Moving towards and integrated approach (Chapter 14). *Water: A shared responsibility,* United Nations World Water Development, Report 2, pp.41-42, 2006.
[12] Nicholls, R.J., Analysis of global impacts of sea-level rise: a case study of flooding. *Physics and Chemistry of the Earth*, Parts A/B/C, 27 (32-34), pp.1455-1466, 2002.
[13] Nicholls, R.J. & Mimura, N., Regional issues raised by sea-level rise and their policy implications. *Climate Research.* Vol.11, pp. 5-18, 1998.
[14] Costello, A. *et al.*, Managing the health effects of climate change. *The Lancet Commission.* Vol. 373, pp.1693-1733, 2009.
[15] Kozka, Z. & Holko, L., Analysis of rainfall-runoff events in the mountain catchment. *Proc. of ERB and Northern European FRIEND Project 5 Conference,* Demänovská dolina, Slovakia, 2002.
[16] Booij, M.J., Impact of climate change on river flooding assessed with different spatial model resolutions. *Journal of Hydrology*, 303 (1-4), pp.176-198, 2005.
[17] Mekong River Commission (MRC), *Overview of the hydrology of the Mekong Basin,* ISSN: 1728: 3248. November, 2005.
[18] Malone, T. *et al.*, Development of improved hydrological forecasting models for the lower Mekong basin. *Proc. Of the 5th AMFF- Improving Inputs Towards Medium Term Flood Forecasting and Warning in the Mekong Basin,* pp.125-131, 2007.
[19] Carroll, D.G., URBS (Unified River Basin Simulator): A rainfall runoff routing model for flood forecasting & design version 4.30, manual, 2007.

[20] Skotner, C., Klinting, A. & Ammentorp, H. C., MIKE Flood Watch-Managing Real-Time Forecasting, DHI Water & Environment, Denmark, 2005.
[21] Narrabri Shire Council, Final Report. *Narrabi flood study review*, Prepared by URC, pp. 11-15, 2009.
[22] Chitprom, P. & Patsinghasaneel, S., Application of hydrodynamic model for flood forecasting in Chao Phraya River Basin for year of 2008, Water Crisis Prevention Center, Department of Water Resources, Thailand, *Proc. of 7th Prince of Songkhla University Eng. Conference*, 21-22 May, 2009.
[23] Taesombat, W. *et al.*, Flood Mitigation Study in Nam Loei River Basin Using River Basin, Modeling. *Proc. of the 3rd National Convention on Water Resources Engineering*, pp.1-10, 2009.
[24] Wannasiri, S., *Rubber Tree,* 3rd edn. Than Kasettakam Press: Bangkok, 1999.
[25] Land Development Department (LDD), Rubber Tree: Technical paper, Office Research and Development of Land Management, Thailand. pp. 11-12, 2005. www.ldd.go.th/Lddwebsite/web_ord/Technical/pdf/P_Technical06020.pdf
[26] South, G. R., Ocean and coastal issues and policy responses in the Pacific islands, Oceans *and Coasts at Rio+10 Assessing Progress, Addressing Continuing, and New Challenges,* UNESCP, Paris, December 3-7, 2001.
[27] Cooper, J.A.G. & McKenna, J., The temporal and spatial dimensions, Social justice in coastal erosion management. *Geoforum,* 39(1), pp.294–306, 2008.
[28] UNEP, *How will global warming affect my world?*, Information Unit for Convention, International Environment House. Switzerland, 2003.
[29] Maneerat, S., *Coastal Erosion in Thailand. Database of marine resources and coastal,* Department of Mineral Resources, Ministry of Natural Resources and Environment, Thailand, 2011 www.km.dmcr.go.th/ index.php?option=com_docman&task=cat_view&gid=124&Itemid=36.
[30] Suvanich, P., *Geohazard, Knowledge and management,* Kampeewan Press: Nonthaburi, 2009.
[31] Jet Propulsion Laboratory, California Institute of Technology, www.flood.firetree.net/?ll=10.5310,99.0967&z=12&m=14&t=3
[32] Rattanamanee, P. *et al.*, Integrated knowledge of shore protection projects, Proc. *of the* 6th Prince of Songkla University Engineering Conference, 8-9 May, 2008.
[33] Thampanya, U. *et al.*, Coastal erosion and mangrove progradation of the Southern Thailand, Journal of Estuarine, *Coastal and Shelf Science*, pp. 68, 75-85, 2006.
[34] Jarupongsakul, T., A future sea-level rise: its impacts on the coastal erosion in the upper Gulf of Thailand, *The Sunday Bangkok Post,* May 21, 2006.
[35] Jarungrattanapong, R. & Manasboonphempool, A., Adaptation strategies for coastal/erosion/flooding: A case study of two communities in Bang Khun Thain District, Bangkok, *TDRI Quarterly Review,* 23(1), pp.11-18, 2008.

Ice dams and backwaters as hydrological risk phenomena – case study: the Bistrita River, upstream of the Izvorul Muntelui Lake (Romania)

G. Romanescu & O. Bounegru
"Alexandru Ioan Cuza" University of Iasi, Romania

Abstract

During the last 20 years, Romanian territory witnessed a recurrence of high-risk hydrologic phenomena. From this point of view, the floods and droughts are particularly noteworthy. Most of the rivers registered historic record levels, while droughts appear even in the mountainous hydrographic networks. Despite the fact that the frequency of rainfall increased during the same time-span, droughts are increasingly present since most of the precipitations are torrential in nature. It is precisely these precipitations of a torrential character that generate flood phenomena. Because Romania harbours a significant number of dams, especially in the mountainous sector, an increase in the number of backwater flooding is noticed. The only phenomenon which manifests itself with some regularity is to be found upstream of the Izvorul Muntelui Lake, on the valley of the River. Various natural and, particularly, anthropic causes favour the formation of ice-dams and Bistrita the flooding of the settlements upstream of the tail of the Izvorul Muntelui Lake. In most cases, the flooding was caused by backwater phenomenon (1–3 m). The overspills are due to ice-dams or the congestion of ice-floes along the course of the Bistrita. The building of houses on the minor bed, and the presence of bridges with the substructure embedded in the mid-channel, have favoured the generation of some of the most spectacular ice-dam ever formed on the river courses of the Carpathians. During winter, a significant contribution to the much-increased discharge levels of the Bistrita River is made by föhn phenomenon present in the upper sector of the river (the Vatra Dornei Depression). Unfortunately, during the last few years, the phenomenon occurs almost yearly, and the damage inflicted is considerable. The

WIT Transactions on Ecology and The Environment, Vol 159, © 2012 WIT Press
www.witpress.com, ISSN 1743-3541 (on-line)
doi:10.2495/FRIAR120141

floods caused by the backwater behind the ice-dams are unique in Romanian territory. The maximum thickness of the ice-dams can reach 6 m.
Keywords: backwater, ice dam, hydrological risk, Romania, water management.

1 Introduction

The dedicated hydrological and climatological literature mentions the onset of the desertification phenomenon in the eastern part of Romania. The last years have witnessed higher average precipitations falling on Romanian territory. Extreme phenomena, such as draughts and floods, are increasingly present, while the precipitations are mostly of a torrential character. The hydrological risk phenomena are primarily due to the massive deforestation taking place in the mountainous sectors [1].

During the last 20 years, most of the major rivers from the eastern part of Romania – the Prut, Siret, Suceava, Trotus, Tazlau, etc. – have witnessed historically-high discharge levels, and the total damage is significant [2–4]. Meanwhile, the droughts, particularly the summer ones, are among the most acute ever recorded in Romanian history (2006, 2011). The available national and international literature on hydrological risk phenomena is very consistent [5–14].

Most of the floods caused by collapsing ice dams occur in the cold (sub-polar and polar) regions of the globe. Those that occur in contexts similar to those of the mountainous regions of Romania are extremely rare. In most cases, they are due to the extensive modification of the environment by human activity. Winter water surges are hydrological risk phenomena which are very rarely observed at medium latitudes and low elevations. They are somewhat frequent in the case of the Mediterranean climate.

Most floods caused by backwater are due to various causes: bridges with under-dimensioned apertures or bridges with piers built on the stream bed; ice jams blocking the watercourses; powerful tides associated with strong winds coming from the sea; the existence of seiches; the presence of dams or dykes which force the excess water upstream in the critical sectors.

The existing Romanian scholarly literature is relatively brief in describing backwater phenomena. The first observation of a backwater flood was made by Dimitrie Cantemir [13] in his famous *Descriptio antiqui et hodierni status Moldaviae*, where he records the gushing of the Danube's waters into the mouths of the Prut River. The most recent Romanian studies mention the backwaters caused by the existence of hydroelectric dams [4], by the ice jams on the rivers from mountainous areas [10], or under marine influence at the mouths of the Danube [11].

The main scope of the present study is to highlight the causes that lead to the formation of ice dams and backwaters, to bring attention to the damage inflicted by the most important winter hydrological risk phenomena, to propose concrete measures for preventing the formation of ice dam and, implicitly, the recurrence of floods.

2 Regional setting

The Bistrita River runs on the central-eastern part of the Eastern Carpathians, and it is the most important right-bank tributary of the Siret River (Figure 1). The Bistrita rises from under the ridge that connects the Gargalau peak with the Cimpoiesul peak from the Rodnei Mountains. The main stream is located at the end facing the Gargalau, at an elevation of 1835 m. In reality, the Bistricioara is the source of the Bistrita, with its exact coordinates at 47°34'27" northern latitude and 24°48'9" eastern longitude. The strongest stream is at the western edge of the Gargalaului, at 1835 m in elevation, while the highest is to the west of it, at 1850 m [4].

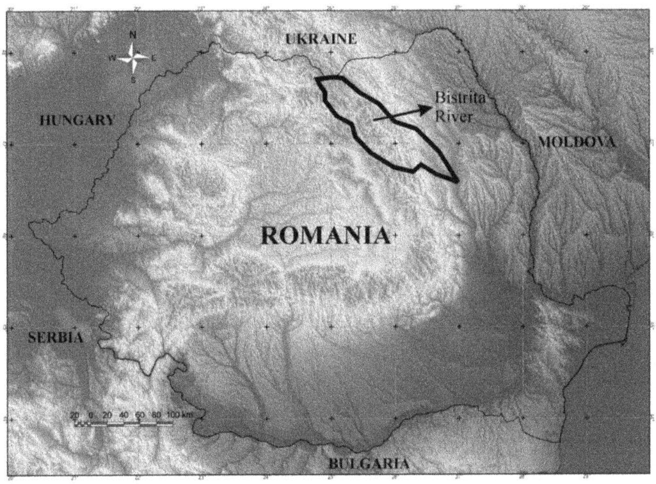

Figure 1: The geographic position of the Bistrita River drainage basin.

It flows into the Siret River, upstream of the city of Bacau, at 46 °29' northern latitude and 26°57'52" eastern longitude. The downstream receiver is the Galbeni reservoir. The elevation of the downstream base level is of 134 m [7, 11].

The Bistrita River has a length of 283 km and collects its waters from an area of 7039 km^2. The average slope is 5‰, the sinuosity index is 1.40, the average elevation is 919 m, and the forested area inside the drainage basin totals 424,301 ha. The river passes through three counties: Suceava, Neamt and Bacau.

The Bistrita has the greatest hydro-energetic potential of all the rivers from the Eastern Carpathians. For this reason, it witnessed intensive hydrotechnical works: along the lower half of its course, no less than 14 hydroelectric power plants were built since the turn of the last century. The most important storage reservoir is the Izvorul Muntelui Lake, built in 1960. This artificial lake is also largely responsible for favouring the development of the ice dams from the Poiana Largului–Farcasa sector.

3 Materials and methods

Since the catchment basin of the Bistrita River is large and its morphology diverse, obtaining the necessary data required a laborious work of collecting from various institutions and conducting field measurements in several representative locations.

Statistical data were obtained from the Department of Water Management from the Siret River, Bacau, from the Moldavian Meteorological Centre, Bacau, and from the Romanian Waters National Administration, Bucharest. The data was processed in the Laboratory of Geo-Archeology within the Faculty of Geography and Geology from the "Alexandru Ioan Cuza" University of Iasi.

Real satellite images were taken from the Remote Sensing and GIS Laboratory of the National Administration of Meteorology and Hydrology, and partially processed or interpreted at the Remote Sensing and GIS Laboratory and the Geo-Archeology Laboratory within the Faculty of Geography and Geology from the "Alexandru Ioan Cuza" University of Iasi. Most of the satellite images that were already processed were taken from the Romanian Space Agency (ROSA) which made them publicly available on the Internet through the PNCD12 Project (http://sigur.rosa.ro).

Land observations and measurements were taken between 2002 and 2003, and in 2009. The major route had flooded river beds Bistrita and Petru Voda. The levels were monitored daily at three stations and topographical measurements were made upstream and downstream from it. All the data analyzed is taken from two meteorological stations (Vatra Dornei and Bicaz) and from three rainfall stations (Carlibaba, Dorna Candreni and Dorna Arini).

The data submitted by the communal and municipal authorities, and the data collected on-site, from local inhabitants or through individual observation, were processed in order to assess the damage at a local and regional level. The data regarding the damage to the infrastructure, especially to roads and railways, was taken from the relevant ministries.

4 Results and discussion

A flash-flood is a sudden and powerful increase of a flowing or static water body's levels. During flash floods, the water overflows its banks, gushing out of the minor bed, and invades the floodplain. Simple flash-floods can be caused by the collapse of a natural dam (an ice dam or a debris dam), heavy rains, snow melting etc. Complex flash floods interfere over the same hydrographical basin and are specific to rivers which flow over diverse geographical regions. The water body which suddenly swells the river and moves downstream is called a flash flood wave. The point of the wave were the elevation (wave height) is greatest is called a crest or peak. Flash floods can be categorised according to size and frequency into: annual flash floods (regular), decennial flash floods (large), exceptional flash floods (once every 100–1000 years; also called catastrophic floods because of their remarkable force and the enormous damage they cause). Retention lakes and dams lessen the impact of the flash floods by trimming their crest and forcing a laminar flow of the water.

In the case of the Bistrita River, of particular interest are the winter flash floods which are not caused by excessive precipitations, but by the factors specific to certain climatic conditions typical of this mountainous riverine sector, as well as by the human-induced changes in the landscape of the Bistrita Valley.

For a better understanding of the processes behind the formation of ice dams and the generation of winter flash floods, we must carefully study the transversal sections of the Bistrita Valley co-jointly with the longitudinal profile of the river. The hydrotechnical facilities, as well as the buildings erected without a preliminary impact study (especially those raised after 1990), should also be considered for our undertaking. Thus, most of the new residences were built on the major bed of the Bistrita, obstructing and practically reducing the real section. Alongside the houses and their outbuildings, small-scale industrial facilities (sawmills) also appeared on the river's major bed. Nonetheless, the most important factor which favours the formation of ice dams is the presence of the Izvorul Muntelui reservoir lake. Its presence facilitated the deposition of alluvial materials at the entrance of the river into the reservoir, and a major decrease of the slope (Figure 2).

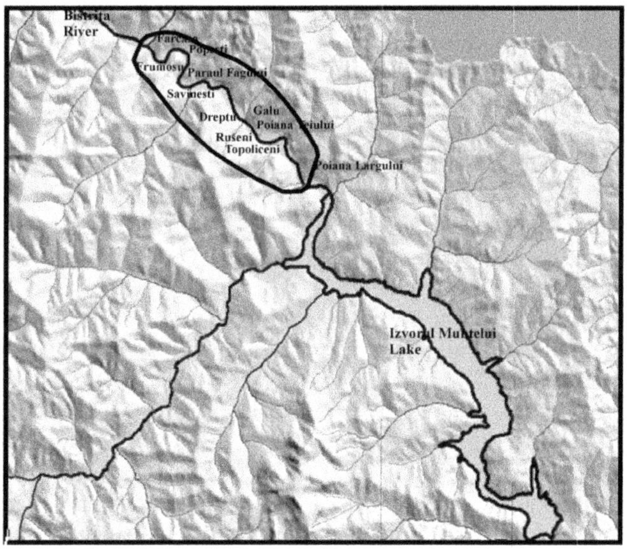

Figure 2: The morphology of the Bistrita Valley and the sector affected by the development of the ice dam.

The Bistrita Valley, upstream of the Izvorul Muntelui Lake, is characterised by a series of narrow sectors which alternate with relatively shallow depression basins, which favoured the establishment of human settlements. Unfortunately, countless numbers of bridges were built to ensure vehicle or pedestrian access in the villages. Because of the decreased span size between the piers of the bridges, the normal release of water out of the ice dams is impeded. The situation is exacerbated by the build-up of logs, branches silt and other debris transported by the previous floods. Most of the tree trunks are due to wood poaching: the heavy logs and thin branches are stored by the poachers on the mountainside, and during torrential rainfall the material is quickly transported into the river bed where they get stuck under the bridges, particularly when the latter have a central pile embedded in the mid-channel. In these cases, an increase of the apertures of between the piles, and even an anthropic sinking of certain sectors of the riverbed, is required to increase the speed of the flowing water, particularly during the winter.

Most of the high-risk hydro-meteorological phenomena occurred, during the last years, between January and April. In the Bistrita Valley, human settlements, defence works, bridges and catwalks, roads, small-scale lumber mills, agricultural lands, etc. have been severely affected.

The causes that can trigger the winter phenomena on the Bistrita river can be grouped into two categories: natural and anthropic.

The natural causes can be synthesised thus: Sustained very-low atmospheric temperatures (< -10°C or even -7°C) for a long time period, which lead to the formation of compact ice formations; Large diurnal temperature fluctuations for short time periods; The successive presence, in the longitudinal profile, of sectors with different slopes, the presence of meanders, and of sectors with narrow widths, including those of the minor bed; The very reduced spans of the Bistrita couloir (the major bed with widths between 300 and 1500 m, and with the minor bed between 50 and 80 m in width); The ingression of air masses with high temperatures, as a consequence of the föhn phenomenon, from the Tihuta Pass towards the Dornelor Depression (the upper basin). The föhn phenomenon causes high temperatures in the upstream sector, which lead to the dislodging of the ice build-up, even during the winter, and its downstream transport and accretion as ice dams. Consequently, the levels of the water rises and the backwater wave causes flooding behind the jams.

The anthropic causes can be grouped thus: The presence of constructions: bridges, houses and outbuildings, fencings, saw mills, etc.; The presence of finalised or non-finalised hydrotechnical works (straight on the river bed): the Topoliceni and Paraul Pantei hydro-electric facilities; The existence of the Izvorul Muntelui reservoir, with direct effects on the hydro-meteorological phenomena due to the alteration of the local base level (the aggradation and avulsion phenomenon through the fan delta depositional feature).

The ice dam is formed in an area reaching 25–30 km in length, up until the settlement of Farcasa (Figure 3). Ever since the Izvorul Muntelui dam was built (it was inaugurated in 1960), this phenomenon occurs regularly, once every 3–5 years. The artificial lake has a length of 33 km and stores a volume of 1.23 billion cubic meters. In 2003, the Topoliceni hydro-energetic facility was brought into service. It is situated upstream of the Izvorul Muntelui reservoir, and its lake has a length of 3.6 km. After this second dam on build on the Bistrita, the problem became more acute: ice dams started to form on a yearly basis, and the maximum thickness of the ice sheet can reach 6–7 m.

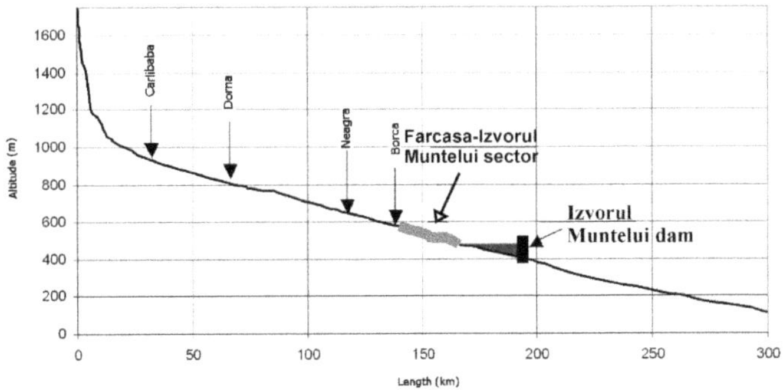

Figure 3: The longitudinal profile of the Bistrita Valley and the outset alignment in the Farcasa–Izvorul Muntelui sector.

In most case, the first occurrences (from December and January) of ice dam happen upstream of the village of Popesti. At the beginning of winter, the thermal inversion phenomenon has not yet manifested, and the moderating character of the Izvorul Muntelui Lake is only felt in the immediate vicinity of it. The entire sector between Farcasa and Poiana Largului is engulfed by the ice dam between February and March, when thermal inversion phenomena take hold of the entire sector between Farcasa and Izvorul Muntelui (Figure 4).

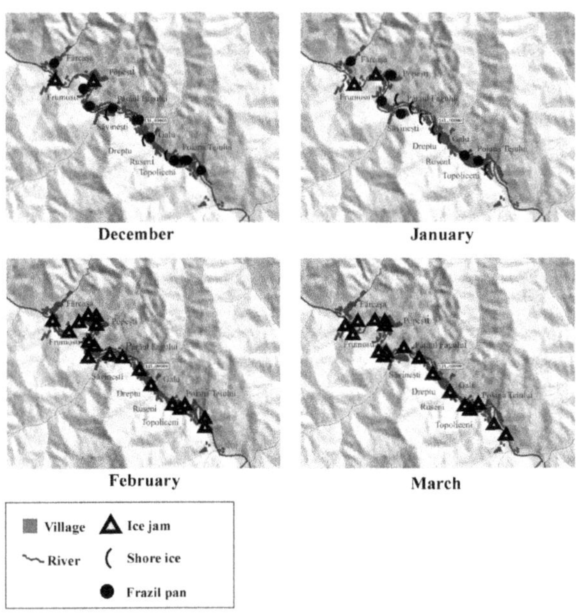

Figure 4: The distribution of the ice dams during the winter of 1995–1996.

During severe winters, with temperatures under -7°C, an ice dam can be in place for the entire time span between December and March (Figure 5). In such cases, the thermal inversions occur as early as December.

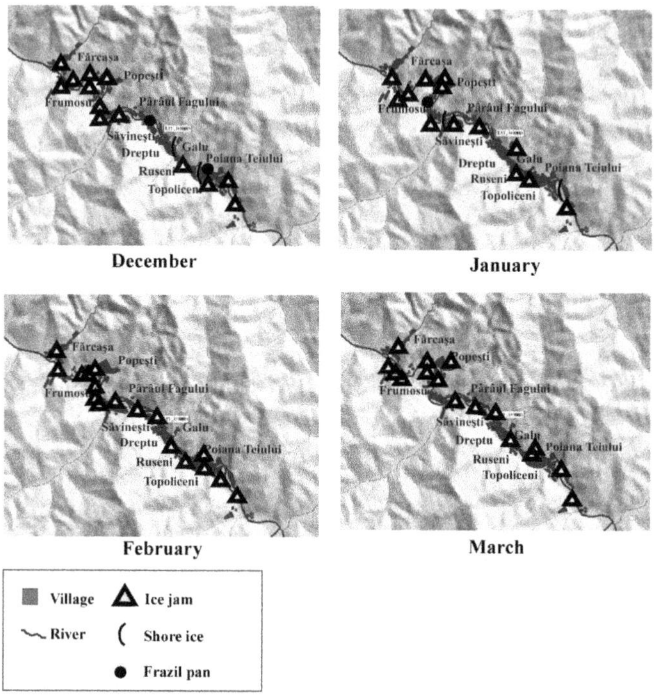

Figure 5: The distribution of ice dams during the winter of 2002–2003.

We should highlight the fact that the ice dams did not start to form immediately after the reservoir appeared in the path of the Bistrita, but only after the silt carried by the river started to accumulate at the tail of the lake. After the delta-fan alluvial cone took shape, and the inclination of the slope was considerably reduced, the first oversized ice dams started to appear (in 1975–1976, 1981–1982, 1998–1999 and 1999–2000). Since the Topoloveni dam was built, the frequency of ice dams increased and, implicitly, so did the frequency of the winter backwaters (in 2000–2001, 2001–2002; 2002–2003, 2003–2004, 2004–2005, 2005–2006, 2006–2007, 2008–2009, 2009–2010, 2010–2011). Consequently, the phenomenon acquired a yearly frequency and it becoming increasingly dangerous.

The most frequent flash floods on Romanian territory occur in spring and at the beginning of summer (April, May, June). Those from during winter are rare and only occur in South-Eastern Romania where some influences of the Sub-Mediterranean climate are observed. The mountainous winter flash floods, from the upper sectors of the Bistrita, are unaffected by the general climate of the area, but by the local climate influences (föhn in the north, and thermal inversion in the south). The föhn is favoured by the intrusion of air masses from the west, through the Tihuta pass, and their descent into the Dornelor Depression. The thermal inversions are favoured by the presence of the depression basins along the Bistrita Valley. In this case, the cold air is dispersed with difficulty, since it stagnates for long periods of time on the bottom of the valley.

In the upper sector of the Bistrita, in the Dornelor Depression (to the north of Farcasa), the dominant winds of the winter months have the highest frequency on the West–East direction (Figure 6). As they descend, they produce the föhn which, in its turn, causes the rapid melting of the snow. This high volume of water spills over the ice from downstream.

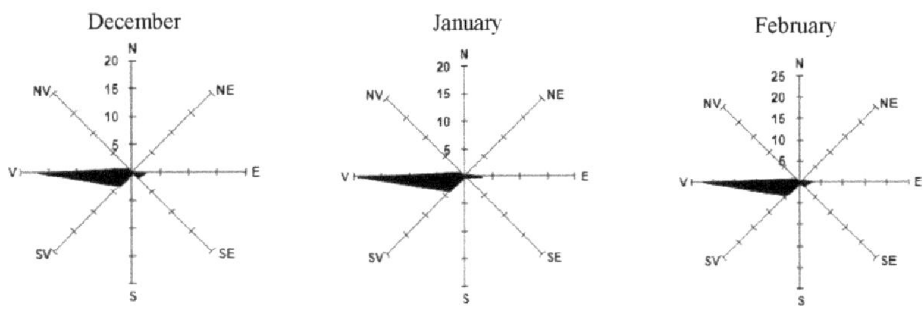

Figure 6: The frequency of the winds according to their directions, during the months of December, January and February 2002 (Vatra Dornei weather station).

The average thickness of the ice dams (or ice jams) is 2–3 m, but they can reach a maximum of 6–7 m. During winters, a peculiar feature of the river is that there is not liquid water present in the minor bed. The waters flow only above the surface of the ice, and in most cases they follow a downstream–upstream path for the 25–30 km long sector from the tail of the Izvorul Muntelui Lake.

The flash-flood which affected the Bistrita River between the 26th of February and the 30th of March, 2002, had the general features similar to the majority of flood phenomena from the Bistrita Valley.

The climatological and hydrological measurements were conducted at the Carlibaba and Dorna Arini (for the Bistrita), and at Dorna Candreni (for the Dorna River, a tributary of the Bistrita). The inflow of water from the Dorna occurred simultaneously with the flash flood on the Bistrita; therefore, the volume of the discharge downstream of the confluence increased considerably. At the confluence, the discharge of the Dorna was much larger than that of the upstream-sector Bistrita. To highlight the role played by the winter 'foehnisation' in the upper sector, several hydrograph charts from the Vatra Dornei–Farcasa sector are herein presented (Figures 7 and 8).

Alongside 'foehnisation', another determining factor is the change of the slope due to the delta-fan alluvial cone formed on the Bistrita at the tail of the Izvorul Muntelui Lake. The alluvial cone caused the aggradation of the flow channel and the avulsion of the river bed.

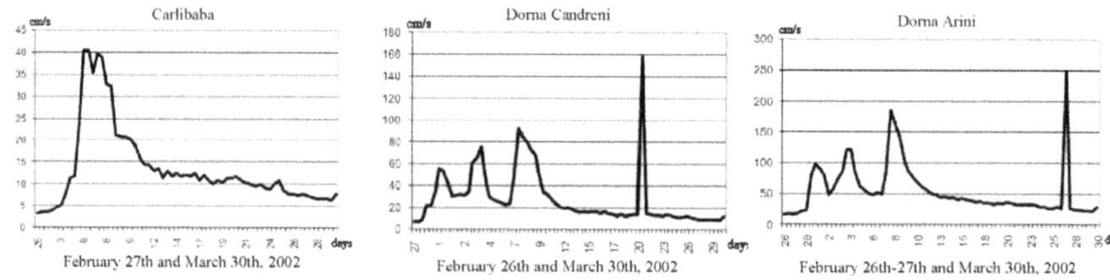

Figure 7: Hydrograph of the flash flood discharge of the Bistrita River at the Carlibaba, Dorna Candreni and Dorna Arini hydrographic stations, between February 26th and March 30th, 2002.

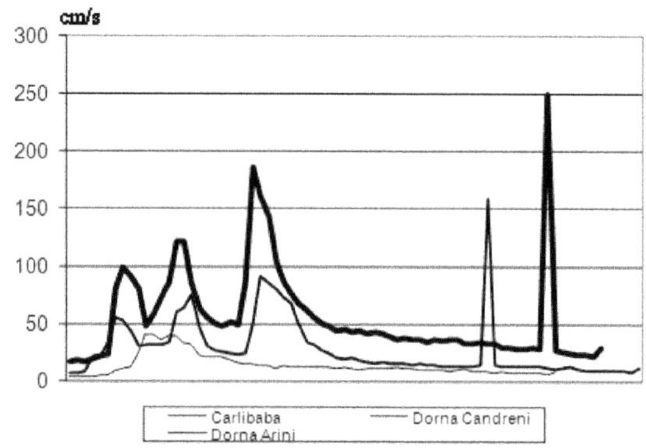

Figure 8: The comparative hydrographs of the flash flood discharges for the Bistrita and Dorna rivers at the Carlibaba, Dorna Candreni and Dorna Arini hydrographic stations, between February 26th-27th and march 30th, 2002.

The alluvial deposits from upstream of the reservoir lake are called "regressive accumulation" in Russian literature, and "progressive aggradation" in American works. This geomorphological feature is the direct result of the forming of fan-deltas and the migration of backwater curve on the river bed, that is to say, the effect of the new local base level [10–11].

During the winter of 2003, a massive accretion of ice-floes created a 7-meter–thick ice plug on the Bistrita. If regular ice dams have an average life of 90–95 days, the one from the winter of 2003 was in good condition for 106 days. The backwater had a maximum intensity between December 31st, 2002, and January 2nd, 2003, during which it raised the water level by 3 m. The greatest material damage occurred during this time; the settlement of Valea Muntelui was hit particularly hard, with 3 people dead and 98 houses damaged. This flood was caused by a warming that took place for a relatively short time span, and during which the temperatures reached 3–5°C.

To prevent further destructive flash floods, particularly upstream of the tail of the Izvorul Muntelui Lake, the following measures must be put into effect: the proper maintenance of the river beds; the sinking of the minor bed as to increase the discharge speed; the foresting of the mountain versants, and the alleviation of soil erosion; the correction of the torrential flow paths through systematisation works; the continuation of the work on the hydro-energetic development of the Bistrita river, in the Borca–Poiana Teiului sector; the regularisation of the river beds, especially in the meandered sectors of the Bistrita, Dorna, and Neagra rivers; the re-calibration of several bridge apertures, and the automatic monitoring of the integrated system.

5 Conclusions

In comparison with other hydrographical systems from the Eastern Carpathians, the Bistrita River also witnesses winter flash floods, due to the combination of several factors: the air 'foehnisation' in the upper sector and the instalment of maximum centres (thermal inversions) in the basins from the lower sectors, which favour the stagnation of the ice dam for long periods of time. The anthropic factor also facilitated the yearly manifestation of the phenomenon.

Although the backwater phenomenon is relatively well known in the Romanian scholarly work, the peculiar features of the phenomenon as it unfolds on the Bistrita River during winters, when massive ice plugs form on its course, make it distinctive and noteworthy. The backwater wave can reach 3 m in height, even when the ice sheet is already 6–7 m thick. To break the ice dams or the ice bridges, the authorities habitually resort to using high-powered explosives (dynamite).

The area affected by these floods is quite large, and the material losses are often significant, but these are rightfully overshadowed by the, albeit occasional, loss of human life. The automatic monitoring of the flash floods and the adequate forewarning of the population is imperative for the future.

Acknowledgements

Our thanks go to the Geo-archaeology Laboratory of the *ARHEOINVEST* Research Platform in the Field of Archaeology, within the Faculty of Geography and Geology from the "Alexandru Ioan Cuza" University of Iasi, which provided the instruments and carried out the processing of the data.

The Ministry of Education and Research paid for the measurements and the publication through CNCS grant no. 0857, for the period 2011–2014, with Professor Octavian Bounegru, Ph. D., as a grant director.

References

[1] Romanescu, G., Siret river basin planning (Romania) and the role of wetlands in diminishing the floods. *WIT Transaction on Ecology and the Environment* **125**, pp. 439-453, 2009. DOI: 10.2495/WRM090391.

[2] Romanescu, G. and Nistor, I., The effect of the July 2005 catastrophic inundations in the Siret River's Lower Watershed, Romania. *Natural Hazards* **57(2)**, pp. 345-368, 2011. DOI: 10.1007/s11069-010-9617-3.

[3] Romanescu, G., Jora, I. and Stoleriu, C., The most important high floods in Vaslui river basin – causes and consequences. *Carpathian Journal of Earth and Environmental Sciences* **6(1)**, pp. 119-132, 2001a.

[4] Romanescu, G., Stoleriu, C. and Romanescu, A.M., Water reservoirs and the risk of accidental flood occurrence. Case study: Stanca–Costesti reservoir and the historical floods of the Prut river in the period July–August 2008, Romania. *Hydrological Processes* **25**, pp. 2056-2070, 2011b. DOI: 10.1002/hyp.7957.

[5] Kuhlicke, C., The dynamics of vulnerability: some preliminary thoughts about the occurrence of radical surprises and a case study on the 2002 flood (Germany). *Natural Hazards* **55(3)**, pp. 671-688, 2010.

[6] Beltaos, S., The role of waves in ice-jam flooding of the Peace-Athabasca Delta. *Hydrological Processes* **21(19)**, pp. 2548-2559, 2007.

[7] Ciaglic, V., Evolutia fenomenului de inghet pe raul Bistrita in iarna 1963-1964. *Hidrotehnica* **10(2)**, pp. 92-101, 1965.

[8] Kowalczyk, T. and Hicks, F., Observations of dynamic ice jam release waves on the Athabasca River near Fort McMurray. *Canadian Journal of Civil Engineering* **34**, pp. 473-484, 2003.

[9] Prowse, T.D. and Beltaos, S., Climatic control of river-ice hydrology: a review. *Hydrological Processes* **16**, pp. 805-822, 2002.

[10] Radoane, M., Ciaglic, V. and Radoane, N., Hydropower impact on the ice jam formation on the upper Bistrita River, Romania. *Cold Regions Science and Technology* 60, pp. 193-204, 2010.

[11] Romanescu, G., Riscul inundatiilor in amonte de lacul Izvorul Muntelui si efectul imediat asupra trasaturilor geomorfologice ale albiei. *Riscuri si catastrofe* **4**, pp. 117-124, 2005.

[12] Shi, J., Zhang, Y. and Zhang, F., Analysis on Backwater Calculation of Ice Jam During Ice-flood Period. *Advances in Materials Research* **243-249**, pp. 4458-4461, 2011.

[13] Cantemir, D., *Descriptio Moldaviae*, Berlin, 1716.

[14] Stefanache, M., *Cercetari privind evolutia unor fenomene hidrologice periculoase*, PhD Thesis, Universitatea Gh.Asachi, Iasi, 2007.

Analysis of runoff variation according to subbasin division

M. Yu, C. Choi, J. Ji & J. Yi
Department of Civil and Transportation Engineering, Ajou University, South Korea

Abstract

Unlike in the past, recently in Korea, heavy rainfall has often been observed out of the rainy season. In terms of patterns, rainfall becomes more focused spatially, and rainfall events occur frequently that exceed the recorded amount of past rainfall by frequency. Consequently, flood disasters happen more often, leading to great economic losses. In this research, flood runoff variation is analyzed according to subbasin divisions. The HEC-HMS model is used to simulate a rainfall-runoff process. The inverse distance weighted (IDW) method is used to distribute the observed rainfall to an entire basin, and the modified Clark (ModClark) method is used to simulate runoff using spatially distributed rainfall and geographic data. Furthermore, the runoff variation pattern is reviewed and compared with the observed data according to the number of subbasin divisions. The results show that the peak runoff increases and the peak time decreases as the number of subbasins increases. However, neither the peak runoff nor the peak time shows these patterns if the number of subbasin divisions exceed nine.

Keywords: HEC-HMS, ModClark, subbasin division, IDW.

1 Introduction

Unlike in the past, in Korea, heavy rainfall has often recently been observed out of the rainy season. In terms of patterns, rainfall has become more focused in spatially, and rainfall events occur frequently that exceed the recorded amount of past rainfall by frequency. All of these tendencies are growing stronger by time. For example, on September 21, 2010, more than 100*mm* an hour of rainfall in the central region of Korea; this caused severe damage, interrupting subway operation and flooding houses. In Seoul, the amount of rainfall was recorded as

WIT Transactions on Ecology and The Environment, Vol 159, © 2012 WIT Press
www.witpress.com, ISSN 1743-3541 (on-line)
doi:10.2495/FRIAR120151

259.2mm in a day, the highest record for late September since official weather observation began in 1908. In the southwestern part of Seoul and Incheon, rainfall at a rate of more than 100mm an hour; the amount of rainfall was 287.5mm in Hwagok-dong, Gangseo-gu, and 86.5mm in Banghak-dong, Dobong-gu, both located in Seoul, showing a significant regional difference. This illustrates that rainfall has become more concentrated in both time and space. As this trend develops, rainfall more often exceeds the recorded amount by frequency, which is calculated from the past observation data. Given this context, it is important to enhance the accuracy of the rainfall-runoff model by reflecting changing flood patterns.

The currently used discharge analysis model by unit hydrograph uses the average rainfall in the basin area as input. This model has the limitation that rainfall data are derived from observation points in particular places. Accordingly, point-based rainfall data are converted to area-based rainfall data to calculate the flooding amount. However, there is a growing need to develop a model that can estimate the spatial distribution of rainfall using point-based rainfall data, especially in places with complex topographies like Korea.

In this study, the inverse distance weighted (IDW) method is applied to utilize rainfall observation data for a broader region, and the modified Clark model [1] is used to simulate rainfall based on spatially distributed rainfall and topography data. In addition, a discharge simulation was conducted according to subbasin division, without modifying the discharge parameter calculated from actual observation data, in order to examine the changing pattern of rainfall discharge according to the number of subbasin divisions. The simulation result is reviewed, analyzed, and compared with the observation data to suggest an optimal method of dividing subbasins.

2 Basic theories

2.1 The modified clark (ModClark) model

The ModClark model adopts the basic principle of the Clark rainfall-runoff model, but adds a simulation function to calculate the discharge volume throughout the basin by adding up the discharge amount of each grid at the basin outlet using the spatially distributed rainfall data and topography data. Fig. 1 illustrates the concepts of the ModClark model [2]. To apply this model, information on the surface area of each cell and the distance between each cell and the basin outlet is needed. Concentration time to basin outlet varies by cell, and is proportionate to the distance. Eqn. (1) shows how to estimate the concentration time to reach the basin outlet:

$$t_{cell} = T_c \frac{d_{cell}}{d_{max}} \tag{1}$$

T_c indicates the entire concentration time to reach the basin, d_{cell} indicates the concentration time between each cell and the basin outlet, and d_{max} indicates the distance between the basin outlet and the farthest grid.

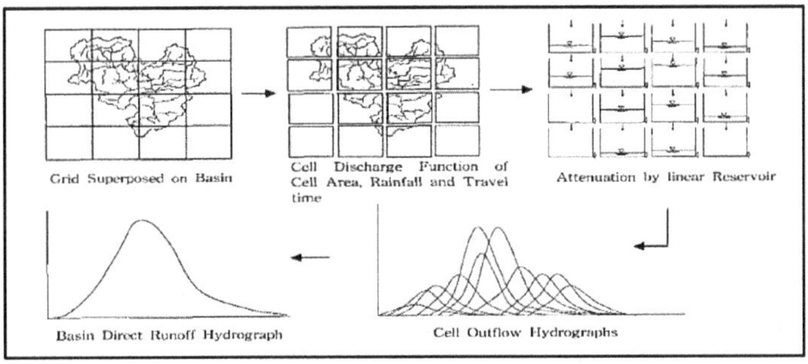

Figure 1: Conceptual diagram of the ModClark model.

Unlike the Clark model, the ModClark model calculates the concentration time for each cell, giving a more accurate estimate of the concentration time from all areas of the basin. It is assumed that the amount of discharge is proportionate to the surface area of the cell. Delayed excess rainfall is tracked by a linear reservoir as follows:

$$O_i = \left[\frac{\Delta t}{K+0.5\Delta t}\right] I_{avg} + \left[1 - \frac{\Delta t}{K+0.5\Delta t}\right] O_{i-1} \tag{2}$$

Here, O_i indicates direct runoff at time i, K is a reservoir constant, and I_{avg} indicates the average inflow between time $i-1$ and i. Δt indicates the time interval. In this study, ArcGIS and HEC-GeoHMS [3] are used to calculate hydrological and topographical data to apply the ModClark method.

2.2 Inverse distance weighted (IDW) method

The IDW method is applied to estimate the spatial rainfall data from point-based observation data. The basic assumption is that two spatially approximate points have similar rainfall data, but the similarity lessens as the distance between the two points grows. Eqn. (3) describes the basic concept of the IDW method:

$$u(x) = \frac{\sum_{i=0}^{N} \omega_i(x) u_i}{\sum_{i=0}^{N} \omega_i(x)} \tag{3}$$

Here, $u(x)$ indicates value to be estimated at point x, N is the number of observation points, $\omega_i(x)$ is the weighted value to be applied to observation point x_i, and u_i represents the observation data at point x_i. $\omega_i(x)$ decreases as the distance becomes greater, as shown in eqn. (4):

$$\omega_i(x) = \frac{1}{d(x,x_i)^p} \tag{4}$$

where $d(x, x_i)$ indicates the distance between observation point x_i and estimate point x. Here, p is the power parameter, that is, the extent of changes in weighted value according to distance; it is also a main parameter of the IDW method.

3 Rainfall-runoff model in test basin

3.1 Test basin

The basin of the Pyeongchang River, the primary branch of the Han River, was selected as a test basin. The Han River basin is located in the central part of the peninsula, covering the area between latitude 36°30′–38°55′N and longitude 126°24′–129°02′E. The basin area is 26,356km^2 (8,455km^2 is in the territory of North Korea); the flow path is 481.7km long, 55.8 km on average; and the basin shape factor is 0.119. It is the largest river in South Korea, covering approximately 23% of the country's territory. The Pyeongchang River basin covers 1,773.39km^2. Its average basin slope is 39.91%, the shape factor is 0.58, and the drainage density is 1.71. The average altitude is EL. 591.53m, while the maximum altitude is EL. 1,570m.

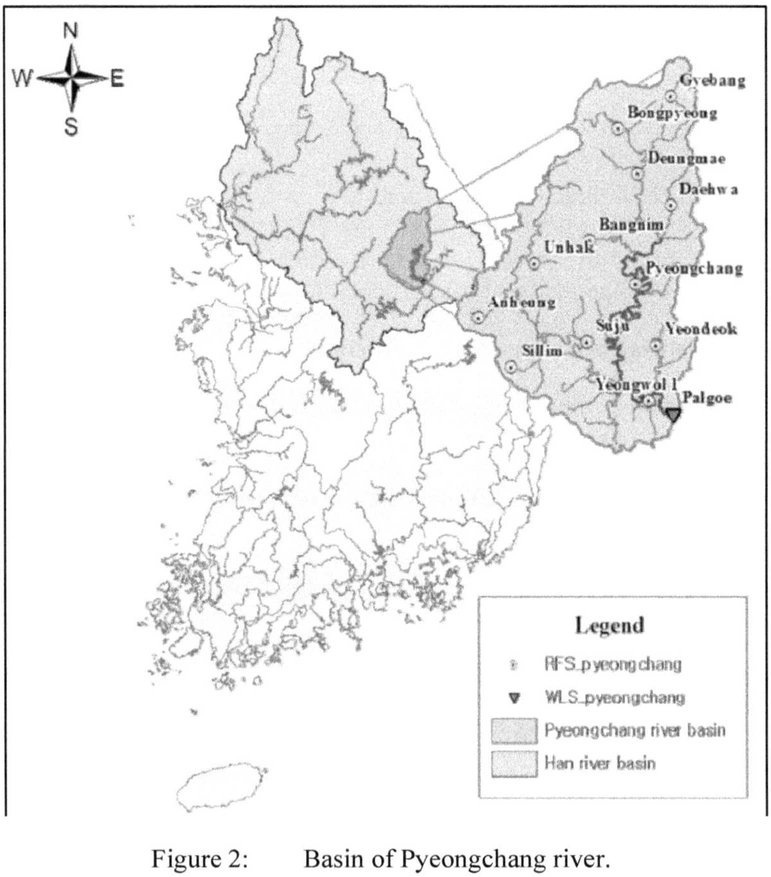

Figure 2: Basin of Pyeongchang river.

3.2 Preliminary treatment

The digital elevation model (DEM) was used to establish the HEC-GeoHMS model, which extracts hydrologic property factors of a basin area based on GIS data. To prevent flow interruption due to lower elevation, repeated calculation was made in the DEM to fill sink areas ("Fill Sinks") until the flow direction could be defined (Fig. 3). Once the flow direction was decided, the number of accumulated flows for each cell could be calculated to map out the pattern of the main river (Fig. 4). Fig. 5 illustrates the formation of the river basin for each factor through river networks built from the accumulated data on flows and flow directions. A detailed soil map and land cover map (Fig. 6 and 7) were used to identify topographical factors and parameters, and the SCS CN coefficient is calculated (Fig. 8).

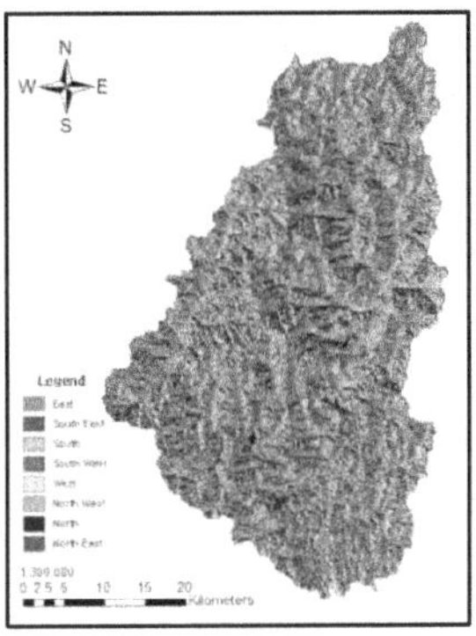

Figure 3: Flow direction.

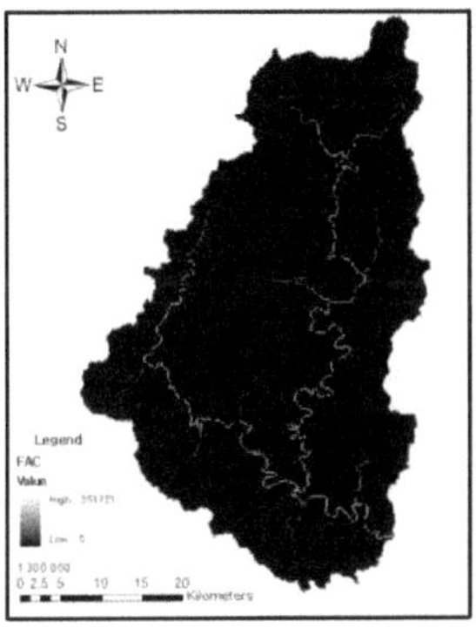

Figure 4: Flow accumulation.

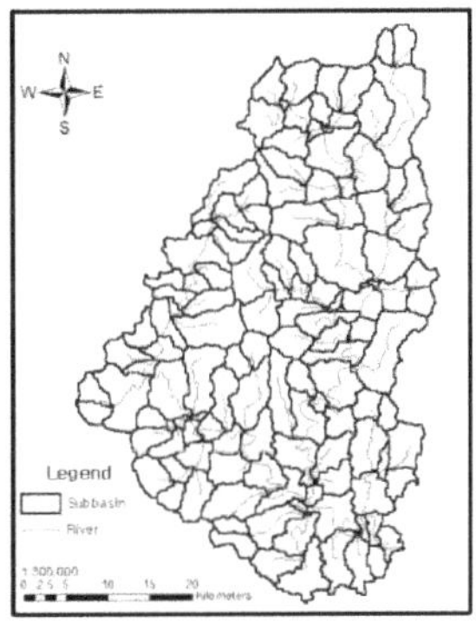

Figure 5: Division of subbasin.

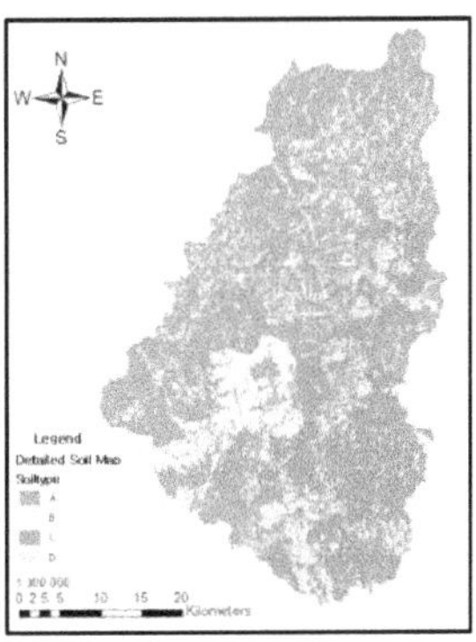

Figure 6: Detailed soil map.

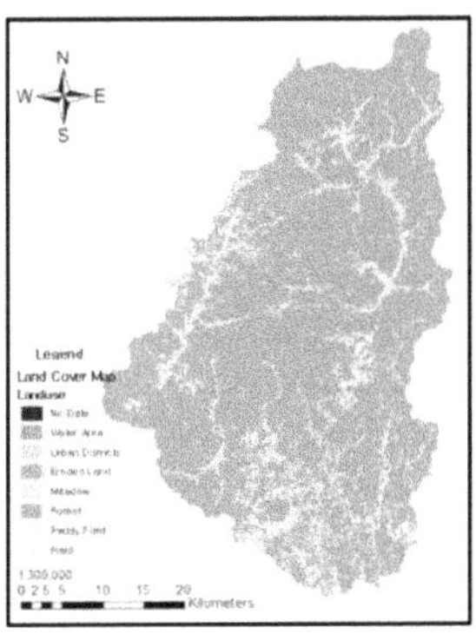

Figure 7: Land cover map.

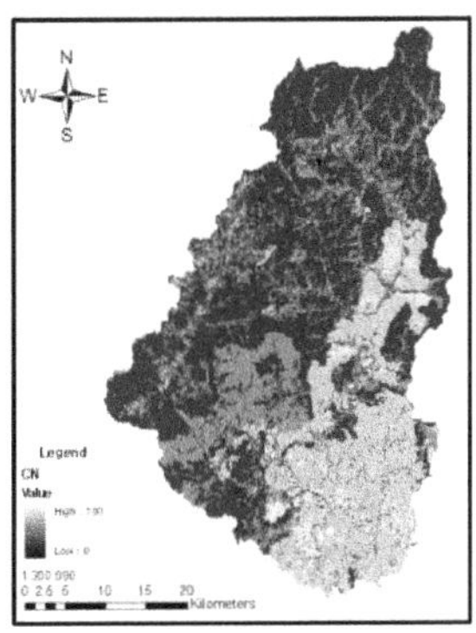

Figure 8: SCS CN.

4 Simulation of rainfall-runoff model by subdivisions

4.1 Subbasin division of the test basin

For simulation, the basin area was divided into nine different numbers of subdivisions: 1, 3, 5, 6, 7, 9, 10, 12 or 15. The subbasins had an equal surface area, and the watershed was set at the confluence point where the Pyeongchang River and the Jucheon river meet, and the Palgye observation point, that is, an outlet of the test basin. The areas of the subdivisions were divided so that they would be as equal as possible, considering the stream order, basin watershed, and so forth. Table 1 lists the areas of the subbasins, their area ratio, flow path (L), concentration time according to the Clark method (T_c), and storage constant (K).

Table 1: Subbasin data and parameters.

Sub-basin	Area (km^2)	L (km)	T_c (hr)	K (hr)	Sub-basin	Area (km^2)	L (km)	T_c (hr)	K (hr)
1-1	1,764.6	134.2	17.8	30.9					
3-1	463.4	63.2	8.4	11.8	3-3	605.5	76.5	10.1	16.3
3-2	695.8	71.0	9.4	11.3					
5-1	273.7	25.3	3.3	2.7	5-4	295.4	33.8	4.5	4.0
5-2	352.8	55.9	7.4	10.7	5-5	310.1	26.6	5.7	6.0
5-3	532.7	53.0	7.0	7.0					
6-1	273.7	25.3	3.3	2.7	6-4	315.7	29.9	4.0	3.3
6-2	236.9	49.1	6.5	11.2	6-5	295.4	33.8	4.5	4.0
6-3	333.0	30.0	4.0	3.2	6-6	310.1	42.8	5.7	6.0
7-1	176.5	22.0	2.9	2.4	7-5	315.7	29.9	4.0	3.3
7-2	239.6	24.3	3.2	2.6	7-6	312.7	28.8	3.8	3.1
7-3	300.7	40.5	5.4	5.4	7-7	202.5	26.6	3.5	3.0
7-4	217.0	23.1	3.1	2.5					
9-1	176.4	22.0	2.9	2.4	9-6	265.4	27.0	3.6	2.9
9-2	218.9	24.3	3.2	2.6	9-7	205.1	12.7	1.7	1.2
9-3	158.4	34.1	4.5	5.5	9-8	186.5	24.6	3.3	2.8
9-4	208.0	23.6	3.1	2.5	9-9	123.5	18.1	2.4	2.0
9-5	222.4	20.3	2.7	2.1					
10-1	154.1	12.0	1.6	1.2	10-6	265.4	27.0	3.6	2.9
10-2	119.6	13.3	1.8	1.3	10-7	146.3	24.4	3.2	2.9
10-3	189.8	38.0	5.0	6.3	10-8	149.1	9.3	1.2	0.9
10-4	209.1	24.9	3.3	2.7	10-9	163.3	19.5	2.6	2.1
10-5	221.2	19.0	2.5	1.9	10-10	146.8	23.3	3.1	2.7
12-1	154.1	12.0	1.6	1.2	12-7	118.2	9.2	1.2	0.9
12-2	119.6	13.3	1.8	1.3	12-8	147.2	9.6	1.3	0.9
12-3	95.3	25.2	3.3	3.8	12-9	146.3	24.4	3.2	2.9
12-4	116.2	18.8	2.5	2.1	12-10	176.2	18.8	2.5	1.9
12-5	187.4	18.9	2.5	1.9	12-11	159.4	15.2	2.0	1.5

Table 1: Continued.

12-6	221.2	19.0	2.5	1.9	12-12	123.5	18.1	2.4	2.0
15-1	154.1	12.0	1.6	1.2	15-9	118.2	27.0	3.6	3.9
15-2	119.6	13.3	1.8	1.3	15-10	147.2	23.7	3.1	2.8
15-3	76.4	15.4	2.0	1.7	15-11	90.3	21.1	2.8	2.7
15-4	69.7	22.5	3.0	3.6	15-12	139.1	8.6	1.1	0.8
15-5	104.5	17.9	2.4	2.0	15-13	163.9	17.0	2.2	1.7
15-6	132.6	7.0	0.9	0.6	15-14	102.0	15.2	2.0	1.6
15-7	102.2	15.9	2.1	1.7	15-15	110.2	14.8	2.0	1.5
15-8	134.7	19.0	2.5	2.0					

4.2 Spatial distribution of rainfall based on the IDW method

To estimate the spatial distribution of rainfall based on data from 12 observation points, IDW was used; this is the most commonly used spatial interpolation method for estimating spatial information based on point observation data. As eqn. (4) shows, the inverse distance squared weighting (IDSW) method was used; this sets the main power parameter at 2 [4].

In HEC-HMS, grid data can only be entered in data storage system (DSS) format [5]. For this study, the hourly rainfall data (in 10min intervals) from the 12 observation points was spatially interpolated and printed out in ASCII file format. It was then converted to a DSS file for data entry. The rainfall data shown on the TM projection of ArcMap can be printed out to illustrate the spatial distribution of rainfall over the entire basin area using each observation point (Fig. 9). The data converted to DSS grid file can be shown in the viewer DSSVue (Fig. 10).

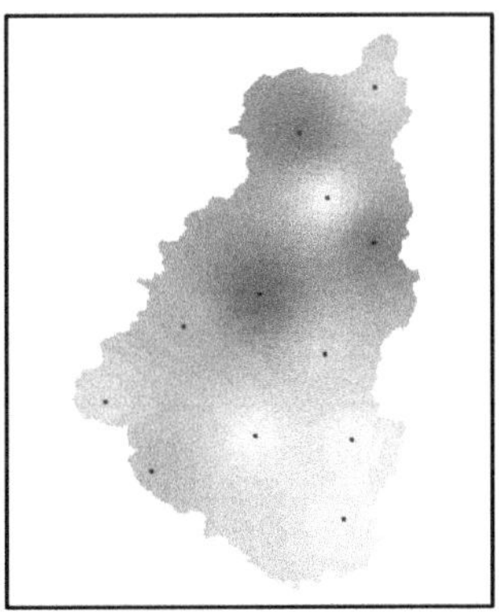

Figure 9: Spatial distribution of rainfall in the TM projection of ArcMap (rainfall data from September 20, 2010).

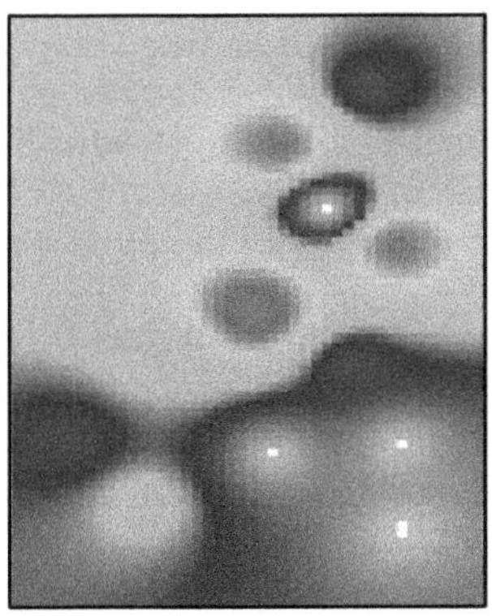

Figure 10: Spatial distribution of rainfall in the DSS grid format (rainfall data from September 20, 2010).

4.3 Simulation result of each rainfall event

Simulation was conducted for different cases of subbasin division for three rainfall events in September 2007, July 2009, and September 2010. The simulation results, including changes in peak flow and peak time by different number of subbasins, were analyzed.

4.3.1 Case one: rainfall event in september 2007

Fig. 11 illustrates the simulation result of rainfall discharge in the subbasins for the rainfall event in September 2007. Table 2 and Fig. 12 show the peak flow estimates (Q_p) in each case, their ratio to the observation data, and the time difference between the peak time estimate (T_p) and observed peak time.

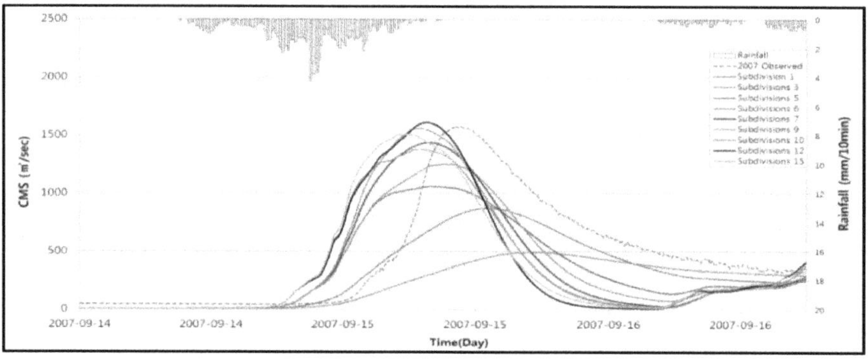

Figure 11: Simulation results by number of subdivisions.

The simulation results show that, in general, the peak flow increases and the peak time is reached earlier as the number of subdivisions increases. The peak flow increased until the number of subdivisions reached nine, and after that, it showed a rather irregular pattern. In terms of peak time, it became earlier until the number of subdivisions was nine—except at five—and reached it earlier except in the case of 12 subdivisions.

Table 2: Changes in peak flow and peak time by number of subdivisions.

Subdivision	Q_p (m^3/s)	Ratio (%)	T_p (Time)	Time Difference (hr)
Observed	1,575.49	-	9/15 15:20	-
Single Basin	492.92	31.3%	9/15 22:40	07:20
Subdivisions 3	873.34	55.4%	9/15 18:10	02:50
Subdivisions 5	1,060.50	67.3%	9/15 13:00	-02:20
Subdivisions 6	1,252.10	79.5%	9/15 14:20	-01:00
Subdivisions 7	1,438.20	91.3%	9/15 12:50	-02:30
Subdivisions 9	1,558.40	98.9%	9/15 12:00	-03:20
Subdivisions 10	1,378.40	87.5%	9/15 12:00	-03:20
Subdivisions 12	1,611.60	102.3%	9/15 12:30	-02:50
Subdivisions 15	1,506.70	95.6%	9/15 10:40	-04:40

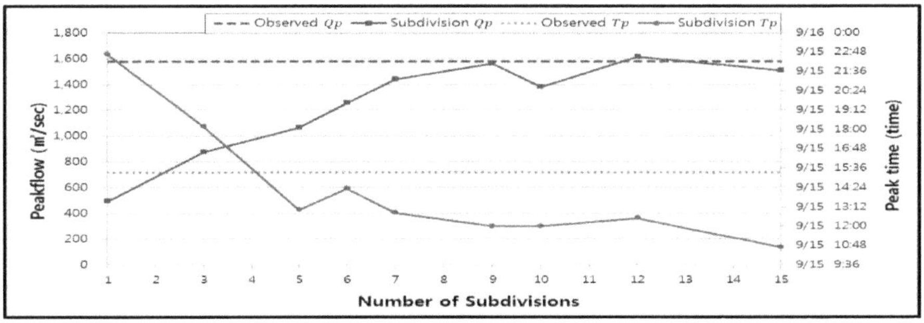

Figure 12: Peak flow and peak time by number of subdivisions.

Compared to the observed outflow hydrograph, the peak flow was closest to the observation data (1,575.5m^3/s) in the case of nine subdivisions (1,558.4m^3/s); as for the peak time, it was most approximate to the observation data (15:20) in the case of six subdivisions (14:20).

Table 3: Changes in peak flow and peak time by number of subdivisions.

Subdivision	Q_p (m^3/s)	Ratio (%)	T_p (Time)	Time Difference (hr)
Observed	1,770.31	-	7/10 02:30	-
Single Basin	502.60	28.4%	7/10 09:40	07:10
Subdivisions 3	904.44	51.1%	7/10 06:20	03:50
Subdivisions 5	1,018.80	57.5%	7/10 04:30	02:00
Subdivisions 6	1,392.90	78.7%	7/10 03:30	01:00
Subdivisions 7	1,418.70	80.1%	7/10 02:30	00:00
Subdivisions 9	1,479.60	83.6%	7/10 01:40	-00:50
Subdivisions 10	1,405.60	79.4%	7/10 01:50	-00:40
Subdivisions 12	1,891.70	106.9%	7/10 01:40	-00:50
Subdivisions 15	1,484.90	83.9%	7/10 00:30	-02:00

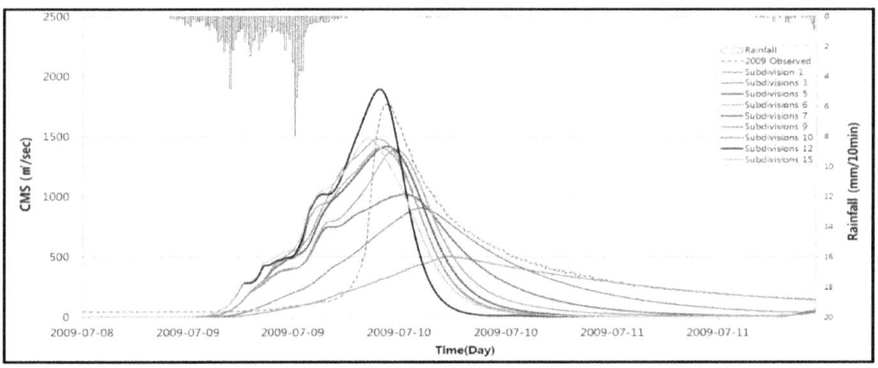

Figure 13: Simulation results by number of subdivisions.

4.3.2 Case two: rainfall event in July 2009

The peak flow increased until the number of subdivisions reached 9, and showed similar volume at numbers of subdivisions between 6 and 15, except at 12. The peak time was reached earlier until the number of subdivisions reached 9, stayed stable until the number was 12, and then became earlier again.

Compared to the observed outflow hydrograph, the peak flow was closest to the observation data (1,770.31m^3/s) in the case of 12 subdivisions (1,891.70m^3/s); the peak time with 7 subdivisions (2:30) coincided with the observation data (02:30).

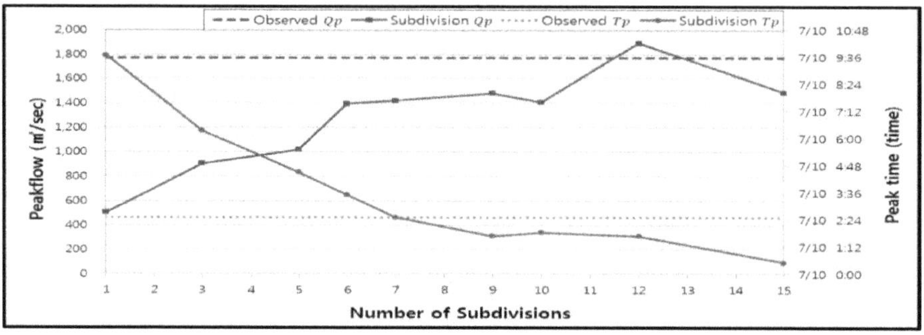

Figure 14: Peak flow and peak time by number of subdivisions.

4.3.3 Case three: rainfall event in September 2010

The peak flow increased until the number of subdivisions reached nine, declined, and then reversed again. Regarding peak time, it was reached earlier until the number of subdivision reached 9, stayed stable until the number reached 10, and then became earlier after the number of subdivisions reached 12.

Compared with the observed outflow hydrograph, the peak flow was closest to the observation data (3,513.7m^3/s) in the case of 15 subdivisions (3,536.9m^3/s), and the peak time in the case of 6 subdivisions (03:20) matched the observation data (03:20).

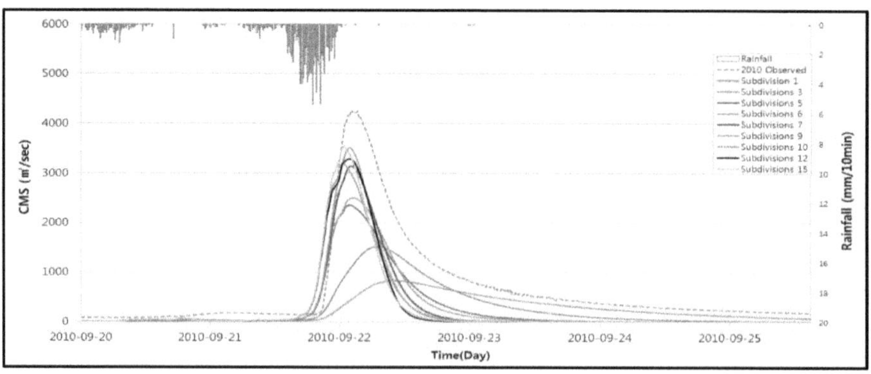

Figure 15: Simulation results by number of subdivisions.

Table 4: Changes in peak flow and peak time by number of subdivisions.

Subdivision	Q_p (m^3/s)	Ratio (%)	T_p (Time)	Time Difference (hr)
Observed	4,249.96	-	9/22 03:20	-
Single Basin	831.01	19.6%	9/22 11:10	07:50
Subdivisions 3	1,518.10	35.7%	9/22 08:00	04:40
Subdivisions 5	2,348.80	55.3%	9/22 02:50	-00:30
Subdivisions 6	2,493.80	58.7%	9/22 03:20	00:00
Subdivisions 7	3,139.90	73.9%	9/22 03:10	-00:10
Subdivisions 9	3,513.70	82.7%	9/22 02:50	-00:30
Subdivisions 10	3,198.70	75.3%	9/22 01:30	-01:50
Subdivisions 12	3,284.20	77.3%	9/22 02:50	-00:30
Subdivisions 15	3,536.90	83.2%	9/22 01:40	-01:40

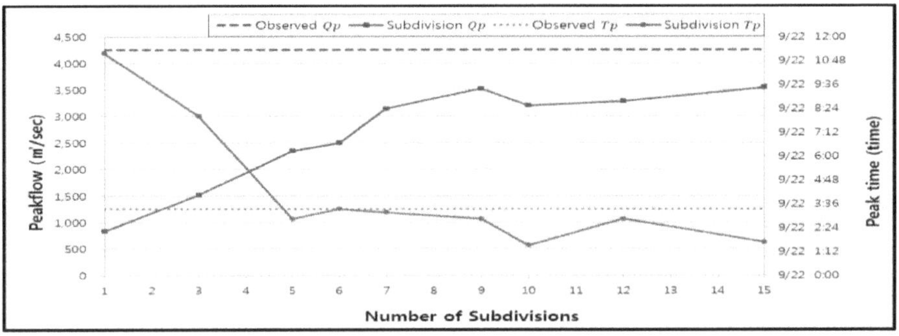

Figure 16: Peak flow and peak time by number of subdivisions.

5 Conclusion

The objectives of this study were to apply the rainfall-runoff model to examine and analyze rainfall discharge patterns according to number of subdivisions in a basin area, and to suggest an optimal way of dividing subdivisions through comparison with the observation data. Accordingly, a simulation was conducted in the Pyeongchang River basin area.

The ModClark model was applied, and rainfall data were processed through the IDW method, estimating the spatial data based on point observation data. The hydrological and topographical factors were identified through GIS. The basin was divided into equal areas of 3, 5, 6, 7, 9, 10, 12 and 15 subdivisions, representing nine cases in total. For each case, rainfall-runoff simulation was conducted for three rainfall events in September 2007, July 2009, and September 2010. To examine the influence of subdivision on peak flow and peak time, the optimization trial of HEC-HMS was not applied.

The simulation results indicated that the peak flows tended to increase and the peak time shifted earlier as the number of subdivisions increased; this tendency weakened after the number of subdivisions reached a certain point. In three rainfall events, comparison between the estimated and the observation data suggested that nine subdivisions (equal area ratio 11.1%) would be an optimal number of subdivisions to estimate peak flow, and six (equal area ratio 16.7%) and seven (equal area ratio 14.3%) subdivisions would be optimal numbers of subdivisions to estimate peak time.

The simulation was limited to the Pyengchang River basin area, and for a broader application of the findings, it will be necessary to conduct additional research and analysis in various settings, including mountainous and flatland basin areas.

Acknowledgements

This study was financially supported by the Construction Technology Innovation Program (08-Tech-Inovation-F01) through the Research Center of Flood Defence Technology for Next Generation in Korea Institute of Construction & Transportation Technology Evaluation and Planning (KICTEP) of Ministry of Land, Transport and Maritime Affairs (MLTM).

References

[1] Kull, D. W., and Feldman, A. D., Evolution of Clark's Unit Graph Method to Spatially Distributed Runoff. *Journal of Hydrologic Engineers*. ASCE, Vol. 3, No.1, pp. 9-19, 1998.

[2] Ahn, S. J., and Yoon, S. H., Runoff Analysis using ModClark Model. *Journal of Korea Water Resources Association*. Vol.38, No.3, pp. 245-257, 2005.

[3] Hydrologic Engineering Center. HEC-GeoHMS: Geospatial Hydrologic Modeling Extension, User's Manual. U.S. Army Corps of Engineers, Davis, CA, 2010.

[4] Donald Shepard, A two-dimensional interpolation function for irregularly-spaced data, *Proceedings of the 1968 ACM National Conference*, pp. 517-524, 1968.

[5] Hydrologic Engineering Center. HEC-HMS: Hydrologic Modeling System, User's Manual. U.S. Army Corps of Engineers, Davis, CA, 2010.

Section 6
Responses to reduce vulnerability to flooding (Special session organised by J. Lamond)

The role of solid waste management as a response to urban flood risk in developing countries, a case study analysis

J. Lamond[1], N. Bhattacharya[2] & R. Bloch[3]
[1]*University of the West of England, Bristol, UK*
[2]*University of Wolverhampton, Wolverhampton, UK*
[3]*GHK International, London, UK*

Abstract

Solid waste management is a large and growing problem for countries in the developing world and is often a neglected aspect of urban management. However, poor waste management can contribute to the impact of urban flooding by blocking drainage, increasing debris and harbouring disease vectors. A review of the literature and analysis of case studies (both from the literature and from examples collected in the preparation of a global urban flood handbook) confirm that solid waste management is an emerging issue in flood risk management practice. Approaches to improve waste management included large municipal programmes and locally based community schemes. It was seen to be important that the management of waste is adopted as part of a wide integrated flood management programme. The research demonstrates that waste management can be an effective response to flood risk but, in order to remain successful, it requires that sufficient commitment and engagement can be mobilised in the long term.
Keywords: waste management, integrated flood management, urban flash flooding, developing countries, community response.

1 Introduction

The management of waste is a perennial problem in developed and developing countries [1–3] and it is an increasingly serious issue in respect of flood risk management [4] for the following main reasons: The poor disposal of waste frequently leads to blockages in drainage and watercourses [5, 6], this effectively

www.witpress.com, ISSN 1743-3541 (on-line)
doi:10.2495/FRIAR120161

reduces their capacity of storage and conveyance and leads to flooding [7, 8]; During a flood waste and other debris collected by floodwaters can cause increased damage to property and lead to higher flood losses [9, 10]; after a flood the deposition of waste can block access and be a source of toxins and breeding ground for disease. Floods can also have an impact on waste management systems [11] leading to leaching of toxins into groundwater. Thus the management of solid waste can be seen as one of the low regret options for reducing flood risk; contributing as it does to quality of life, health and development generally, while also lowering the impact of disasters [12, 13]. Low regret options are important for flood management in developing countries in particular because financial pressures render the integration of flood management with wider development goals especially critical.

Low priority is often accorded to the management of solid waste in developing countries for a number of reasons: waste management is frequently seen as a low status occupation with poor wages and therefore there may be a lack of relevant training and expertise [14]; low awareness of the health and sanitation implications of poor waste management can lead to an absence of pressure from civil society and therefore funds are diverted to other more high profile programmes; this lack of resources perpetuates the low status of waste management with no money to purchase equipment, train staff and develop disposal sites [14]. Poor wages can also encourage absenteeism. These factors not only result in inadequate collection and disposal of waste, they also tend to lead to concentration of resources in city centres and total neglect of the needs of informal settlements [4].

This paper therefore considers the role and awareness of waste management in flood risk reduction by reviewing the literature related to flood risk management and solid waste management. The review was focussed in particular on case studies that demonstrate the combination of waste management and flood risk reduction in developing countries. It aimed to address three research questions:

1. Is poor solid waste management a major contributor to flood risk management?
2. Is solid waste management an effective measure in flood risk reduction?
3. Are community based measures successful in improving the disposal of solid waste and reducing flooding?

2 Literature review

The method employed in the analysis is review of literature relating to case studies of flood risk involving poor solid waste management in the developing world. Case studies were identified from examples collected in the course of the preparation of an urban flood risk management handbook for the World Bank [14], from UN Habitat case studies [4] and from a general keyword search. More case studies were identified than could be included in the scope of this analysis, therefore case studies were selected on the basis of highest relevance to the research questions while ensuring examples from Asia, Latin America and Africa were included.

Solid waste management forms part of a complex urban management system that is interlinked [15] and frequently underfunded. This leads to the combination of poorly constructed and maintained drainage systems with inadequate waste management. These issues are not confined to the developing world [15] but they are more acutely felt in emerging economies where the rapid growth in urban populations are likely to result in massive increases in the demand for waste disposal and the breakdown, due to overloading, of existing functioning systems [15].

As noted above, in neighbourhoods with inadequate solid-waste management or drain maintenance, waste and vegetation can quickly clog conveyance systems and drains, leading to flooding. In particular solid waste can have an impact on the incidence of localized flooding with light rainfall. For instance, in Georgetown, Guyana this led to 29 local floods between 1990 and 1996 [16]. Residents in Tanzania also recognise that poor solid waste management blocking streams has exacerbated their flood risk [17].

The type and quantity of waste generated differs significantly between countries and within urban areas [5]. Industrialized countries and urban areas will tend to generate greater quantities of waste compared to developing countries and rural areas. In Asia the amount ranges from 0.5 kilogram per person per day in Nepal to more than 5 kilograms per person per day in Hong Kong [3]. In Latin America the range is said to be 0.5-1.0 kilogram/day per person [18]. Disposal and collection capacity also varies, for instance in cities like La-Paz, and Brasilia the total collection of solid waste is up to 90% on the other hand in Santiago the total collection is less than 57% [18]. In Kumasi, Ghana, 40% of waste is said to end up in unauthorised garbage dumps and often the river [8]. Waste from upstream areas can also collect in cities as for example in Port-au-Prince in Haiti [6] and Managua in Nicaragua [14].

Waste management is comprised of three main steps: collection, transportation and disposal. Collection can be further divided into primary, secondary and tertiary stages reflecting the scale of the operation and resulting in the amalgamation of multiple waste streams into a body of waste materials that require disposal. The hierarchy of reduce, reuse, recycle, energy from waste disposal can be applied before collection, at the point of collection and during disposal to minimise the amount of waste eventually consigned to landfill. It has been observed that the engagement of populations in the process of waste reduction can be a highly effective mechanism to control waste. Therefore raising the awareness of residents regarding the benefits of waste reduction can be a first step in responsible waste management and flood risk reduction [2]. Community involvement and participation in community based approaches to the management of solid waste and improvement of drainage are often advocated. Lack of public participation is said to result in wasted municipal effort as, without understanding and awareness, locals will continue to dump their waste in drains [13].

3 Case studies

3.1 Accra (Ghana)

The 2007 floods in Accra, the capital city of Ghana, caused hundreds of people to leave their houses and move to safer locations [19]. Similar situations occurred in 1995, 1999, 2001, 2008 and 2011 causing havoc. Local flooding in Ghana is often attributed to blocked drainage caused by poor waste management [20]. However populations in Accra are hardly aware of the drainage issues associated with flooding and regard waste management as the role of the government [2]. It is observed that the population are not resilient and adaptive to the existing risk [21] and the level of preparedness, recovery and response is very low.

3.2 Mumbai

Mumbai in India is subject to many sources of flood risk including increased impervious areas and loss of storage within the city. Poor storm water drainage is, nevertheless, a major contributory factor to the severity of regular monsoon flooding [22]. Indiscriminate dumping of solid waste combined with storm water and municipal wastewater often clogs the drainage system resulting in coastal flooding and inundation during monsoon months [23]. In response to the 2005 flood the state put in place a ban on the sale and use of plastic bags. However this has not yet been effective due to lack of awareness and enforcement [1].

3.3 Jakarta

Jakarta, the capital of Indonesia, an urban metropolis of 27 million inhabitants is subject to the risk of flooding from the sea, rivers and intense rainfall. Subsidence of the city is increasing the risk to Jakarta and massive urbanisation has led to reduction of storage areas for water throughout the city. Major channelization of drainage through and around the city was designed to convey flows representing the 25 year flood event. However, silting and poor waste management has resulted in the reduction of the channel capacity to a one in five year level. Although waste collection was privatised in some districts of Jakarta in the late 1980s [24], this initiative covered only middle to high income residents and the slow disposal of collected waste still results in large rubbish dumps in informal areas [25]. Local community based drain clearance, waste recycling and removal schemes have been successful in generating employment and raising awareness of the importance of drainage, but the scale of the urban waste stream and the carriage of waste from upstream into the city means that the local efforts are not effective in reducing flood risk [26].

3.4 Bamako

Bamako, the capital of Mali, has a population of approximately one million with roughly half residing in informal settlements. In 1999 flash flooding throughout

Bamako caused death and destruction and analysis showed that poor waste disposal was a significant contributory factor. Consequently a four year project was aimed at improving the management of storm water and solid waste in one of the worst affected districts of the city. Among other things the project set up improved waste collection: eight collection routes, each served by a collection team using tractor-trailers, with disposal at a nearby landfill. The scheme also created local employment for refuse collectors. Stakeholder participation, combined with a comprehensive planning framework, was used for the first time in the city. This helped to build consensus and increase capacity within Non Governmental Organisations and Community based organisations. Channel volume was restored and flood risk reduced and Bamako has not experienced a similar flood disaster since 1999, partly as a result of these measures [27]. However, the City of Bamako in general still faces big challenges in waste disposal as responsibility for funding activities is not clear and the final disposal sites are inadequate. The planned landfill site 30km outside the city has not yet been constructed [4].

3.5 Maputo

Maputo, the capital of Mozambique, houses 45% of the total Mozambican urban population, 50% of which is considered to live below the poverty line. During the 2000 flood 70 percent of flood deaths were in urban areas near Maputo, mainly in the cities of Xai-Xai and Chokwe [28]. Maputo is highly vulnerable to flooding from sea level rise, fluvial and pluvial flooding. In Maputo risk has been reduced by construction of improved drainage systems on the border of the Mafala district. The introduction of a "garbage tax" to finance improvements in solid waste management was administered. The government negotiated public service contracts that institutionalized the primary collection as a free-of-charge public service for all residents [29]. However, despite this investment, during the 2010 floods solid waste blocked the new channels and emergency clearance was necessary. Flood risk mitigation efforts in Mozambique now include many non-structural measures such as preparedness education that stresses the need to keep drainage clear [14].

3.6 Marikina

In the Philippines, regeneration of the Marikina River was undertaken, to enhance the local area and reduce flooding. After clearing the river the municipality introduced a strict waste collection policy which forbade dumping of garbage outside of houses except on collection days. The policy is enforced via penalties for non-compliance. The river is much more attractive as a result and flooding has been reduced, however success of this policy is limited by the influx of waste from settlements further upstream [30].

3.7 Cotonou

Cotonou with a population of about 700,000 is Benin's largest urban centre and major port. Large areas of commercial and residential importance are at risk from sea-level rise and storm surges. In addition the drainage system is inadequate and poor solid waste management contributes to drainage problems, worsening the impact of flooding [31]. Within the informal settlements waste management is not provided by the municipal authority, centralisation has ensured that waste collection is concentrated in the city centre. In the 1990s Pr A.P.E. The "Programme Assainissement et Protection de l'Environment" commenced in the Sainte-Rita municipality. This participatory programme trained and employed youths and women to collect and recycle waste and a nascent private sector arose backed by a community bank [4]. Unfortunately many of these firms subsequently went out of business because of the low fees.

3.8 Lagos

Lagos is a megacity in Nigeria, West Africa of about 10 million inhabitants. The city is low lying and highly prone to flood from sea-level rise and storm surges. Apart from that in the rainy season, intense rainfall results in flooding as matter of course. Drainage systems are in general inadequate and often blocked due to poor waste management practice. Interviews in the informal settlements near the coast revealed that flooding was one of the most serious problems they faced and confirmed that poor waste management and clogged drains were one of the major causes of the flooding [32].

Awareness of solid waste issues was seen to be high in Lagos, but in 2008 there were few community initiatives to clear blocked channels. Now there is a city-wide strategy in place and Lagos has recently been subject to a greatly improved waste management programme [33] including street advocacy and highly visible workers to create awareness.

3.9 Managua

Flooding in Managua, the capital of Nicaragua, is greatly increased by the blocking of the many watercourses that run through this dispersed urban area. The population of approximately two million [34] is sited on steep rugged terrain and subject to earthquakes and flooding. The recent population growth has led to deficits in basic facilities and services and weak urban governance resulting in serious shortcomings in the refuse collection service [1]. To address the issue in 1980 a system of micro-dams was established with the dual purpose of attenuating floods and retaining refuse. Over 16 dams have been constructed Figures show that they extract in excess of $500m^3$ of sediment from the river system [34]. However, flooding is still a yearly event in several neighbourhoods of Managua due to the lack of a drainage master plan and poor urban planning. The lack of flood awareness means that settlements still grow into the most risky areas. Recent initiatives in solid waste management [1] may lead to improved flood risk, athough the driving force behind the initiatives is not flood reduction.

3.10 Mexico City

Mexico City, Mexico has been subject to greatly increased incidence of flash flooding over the 20th century. The high urban density ensures that intense rainfall events, such as occurred in 2006, cause massive flooding. Informal settlements are particularly vulnerable. The area is at risk of multiple hazards as steep terrain can result in landslides and loss of life and the accumulation of trash and debris in the inadequate drainage system worsen the impacts and make recovery slower [36].

3.11 Georgetown

In Georgetown, Guyana, an assessment of the drivers for vulnerability of neighbourhoods to flooding evaluated the main causes in four wards [16]. Drainage issues were identified by residents of all four wards because drainage channels were blocked by a combination of overgrowth, silt and waste. The study concluded that the clearance of drainage by residents in one ward had contributed to the reduction of flooding in this area. In other areas lack of municipal waste collection and no community action had resulted in continued flooding. Residents of these areas took other approaches to minimise damage such as moving out of the ground floor.

4 Results

The case studies (summarised in table 1) reinforced existing perceptions in the flood management literature regarding the importance of solid waste management in reducing flood risk. However these examples also highlighted the fact that no single approach to waste management was universally successful in practice.

Literature demonstrates that poor waste management is a perennial problem that is set to get worse with the rapid expansion of urban areas in the developing world and the inability of municipalities to resource the necessary waste collection. The case studies confirmed that solid waste management is a major contributory factor to urban flooding across the globe with examples from Asia, Africa and Latin America. For many of the examples blocked drainage is particularly implicated in localised flash flooding; but solid waste was also seen to worsen the impact of all types of floods.

The case studies illustrate the fact that the impact of poor solid waste management is experienced very keenly by residents of informal settlements that are often outside the control of municipal authorities. These residents are also more vulnerable to the impacts of flooding and less able to cope with it: therefore improvements in solid waste management can have a disproportionate effect on reducing vulnerability.

In several of the case studies informal settlements were not served by municipal authority collection and disposal schemes. In such settlements a combination of private enterprise and community participation was seen to be

Table 1: Summary of main features of case studies.

Case study	Problem identified	Community awareness	Municipal action	Community action
Bamako	Poor waste management a major factor in 1999 flood	Awareness was raised during the programme	Structural drainage improvement was carried out	Local waste collection system set up. Disposal still a problem
Accra	Blockage of drainage causes flooding	Residents not aware	Authorities appear overwhelmed	Residents regard it as the role of the government
Cotonou	Indiscriminate dumping of waste	Yes and some have taken action	Peripheral areas neglected as the focus is on the city centre	Locals trained to collect and gain revenue from waste
Maputo	Flooding caused by inadequate drainage in the city	Awareness low but being targeted by education programmes	Some urban drainage improvement programmes in Maputo	Communities still dump rubbish in drains, causing major problems in 2010 flood
Lagos	Flooding due to blocked drainage	High awareness in community	New city wide strategy of waste disposal	Very few community initiatives
Marikina	Flooding partly due to waste clogging the river	Residents made aware by the programme	River dredging and penalties for dumping	Residents comply but waste still arrives from further upstream
Jakarta	Blocked channels cause widespread flooding	Residents aware of the problem	Government plans to dredge channels but hampered by informal settlements	Local community based schemes successful but waste is carried from upstream areas
Mumbai	Plastic bags blamed for flooding	Poor awareness	Ban on use of plastic bags	Poor compliance with ban
Guyana	Clogged and inadequate drainage leading to flooding	Identified by residents as important.	Authorities have other priorities	In one ward CBO had cleared drains, this ward had not flooded
Mexico City	Waste blocks drains and leads to flash flooding	Appears to be low awareness	Programmes focus on other priorities	No evidence of community action
Managua	Waste in the rivers worsen flooding	Awareness is poor	Network of micro dams to collect rubbish and silt	Low compliance, flooding still an annual event

the only way forward to improve waste management. Communities can generate employment opportunities and revenue by sorting and recycling waste rather than dumping it.

Another feature of the case studies is that interventions in drainage and solid waste by governments and NGOs were often effective in the short term but their effectiveness reduced in the longer term as the impetus faded and systems were not maintained. This tendency for systems to fall into disuse may be partly compounded by the fact, also seen in the examples here, that locally based schemes cannot wholly address the blocked drainage problems, as new waste is conveyed from upstream areas that are out of the control of the local activists.

5 Conclusions

1. Poor solid waste management contributes to urban flooding across the globe and, as urban populations increase, is likely to be an increasingly important factor.
2. The control of solid waste within areas at risk of flooding has the potential to reduce risk by minimising the amount of waste blocking drainage channels. However local programmes are limited in their effectiveness as solid waste is often carried downstream and residents do not reap the benefit of reducing their own waste.
3. Reduced accumulation of waste in general will reduce the environmental impact of flooding and has multiple other benefits to health and wellbeing.
4. Engagement of communities in risk reduction is helpful in increasing resources and in generating the necessary awareness and motivation to reduce dumping. Measures implemented without community participation are often undermined by poor waste disposal habits.
5. There is a need to educate the public to bring change in their attitude towards dumping and reduction of waste. Disapproval of undesirable activities and tackling apathy towards appropriate operation and maintenance of existing infrastructure may bring forth good user practices.
6. Community based measures need to be a part of a wider waste management plan in order to make a permanent scalable difference to the amount of waste dumped in cities and by extension to flood risk.
7. Flood risk reduction via solid waste management can be effective as part of an integrated flood management plan, however, it requires long term, ongoing, commitment from municipalities and communities.

References

[1] UN Habitat, Solid waste management in the world's cities, Water and Sanitation in the Worlds Cities 2010, London/Washington: Earthscan, 2010.

[2] Sam. P.A. Are the Municipal Solid Waste Management Practices Causing Flooding During the Rainy Season in Accra, Ghana, West Africa, online [accessed 09/05/2011], Modern Ghana. 2009.

[3] Hoornweg, D., and Thomas, L. *What a waste: solid waste management in Asia*. IN REGION, E. A. A. P. (Ed.) Urban Development sector unit working paper series no. 1. Washington DC, The World Bank.1999.

[4] UN HABITAT, The State of African Cities 2010: Governance, Inequality and Urban Land Markets. Nairobi, UN Habitat. 2010.

[5] Muñoz-Cadena, C.E., Arenas-Huertero, F.J. and Ramón-Gallegos, E. Comparative analysis of the street generation of inorganic urban solid waste (IUSW) in two neighborhoods of Mexico City, *Waste Management*, **29(3)**: p.1167-1175. 2009.

[6] Bras, A., Berdier, C., Emmanuel, E. and Zimmerman, M., Problems and current practices of solid waste management in Port-au-Prince (Haiti), *Waste Management*, **29(11):** p. 2907-2909. 2009.

[7] Marfai, M.A., *Urban coastal flooding and risk management (an overview of Jakarta and Semarang City.* Presentation in the consultation workshop *Preparation of a Global Handbook for Urban Flood Risk Management,* Jakarta, May 25-6. 2011.

[8] Appiah, D.O. The Dilemma of Poverty and Safety: The Case of Urban Flooding in the Aboabo River Basin, In Jha, A. Lamond, J., and Bloch, R. Cities and Flooding: A Guide to Integrated Urban Flood Risk Management for the 21st Century, Washington, GFDRR/World Bank. 2012.

[9] Nicholas, J. and Proverbs, D. Benchmarking the assessment of flood damaged domestic properties. London. Royal Institute of Chartered Surveyors. 2002.

[10] Chen, J., Tsai, H., Hsu, P. and Shen, C. 2007, Estimation of waste generation from floods, *Waste Management.* **27(12):** p. 1717-1724. 2007.

[11] Pilapitiya, S., Vidanaarachchi, C. and Yuen, S., Effects of the tsunami on waste management in Sri Lanka, Waste Management, **26(2):** p. 107-109. 2006.

[12] Jha, A. Lamond, J., and Bloch, R. Cities and Flooding: A Guide to Integrated Urban Flood Risk Management for the 21st Century, Washington, GFDRR/World Bank. 2012.

[13] ADPC. Mainstreaming disaster risk reduction, Urban Governance and Community Resilience Guides, no 4. ADPC, 2010.

[14] Jha, A., Lamond, J., Bhattacharya, N., Lopez, A., Bird, A., Proverbs, D., Davies, J., Papachristodolou, N. and Bloch, R. Five feet high and rising, Cities and Floods in the 21st Century. Policy Research Working Paper Series 5648. Washington DC, World Bank. 2011.

[15] Ruth, M., Coelho, D., Understanding and managing the complexity of urban systems under climate change, *Climate Policy* **7(4)**: p. 317–336, 2007.

[16] Pelling M., What determines vulnerability to floods: a case study in Georgetown, Guyana, *Environment and Urbanization*, **9(1):** p. 203–226. 1997.

[17] Dodman, D., Kibona, E., and Kiluma, L., Tomorrow is too late: Responding to Social and Climate Vulnerability in Dar es Salaam, Tanzania. Case study prepared for Cities and Climate Change: Global Report on Human Settlements 2011, UN-habitat. 2011.

[18] USAID. Environmental issues and best practices for solid waste management, Environmental guidelines for the USAID Latin America and Caribbean Bureau. 2006.
[19] Concern universal. Concern Universal Flood Relief in Northern Ghana – September 2007. http://www.concernuniversal.org/index.php?/where_we_work/ghana/flood_relief_in_north_ghana,_september_2007. Date accessed 19.04.2011. 2007.
[20] Boadi K, Kuitunene M. Municipal Solid Waste Management in the Accra Metropolitan Area, Ghana. *The Environmentalist*. 2003; 23: p. 211-218, 2003.
[21] Ahadzie, D. K., and Proverbs, D. G. Flooding and Post Flooding Response Strategies in Ghana. *WIT Transactions on Ecological Environment*, 33: p. 281-291, 2010.
[22] Gupta, K., Urban flood resilience planning and management and lessons for the future: a case study of Mumbai, India, *Urban Water Journal*, **4(3):** p. 183-194. 2007.
[23] Murthy, R. C., Rao, Y. R. and Inamdar, A. B. Integrated coastal management of Mumbai metropolitan region, *Ocean & Coastal Management,* 44: p.355–369. 2001.
[24] Klundert, A., and Lardinois, I. Community and Private (formal and informal) sector involvement in municipal solid waste management in developing countries, Background Paper presented at the Ittingen Workshop, Ittingen, Switzerland April 10/12. 1995.
[25] Cointreau-Levine, S. *Private sector participation in municipal solid waste services in developing countries*. Vol. 1. The formal sector. Urban Management Programme Discussion Paper No. 13. World Bank. Washington, United States, 1994.
[26] Mercycorps, *Community Based Flood Risk Management Initiatives. Learning from flood risk reduction projects in* Jakarta Presentation in the consultation workshop *Preparation of a Global Handbook for Urban Flood Risk Management,* Jakarta, May 25-6. 2011.
[27] Setchell, C. A., Multi-sector disaster risk reduction as a sustainable development template, the Bamako flood hazard mitigation project. *Monday Developments*, Interaction. 2008.
[28] UN-HABITAT Bamako – Using Partnerships to Support Environmental Management. http://ww2.unhabitat.org/programs/uef/cities/summary/bamako.htm. 2007.
[29] Kruks Wisner, G. *After the Flood: Crisis, Voice and Innovation in Maputo's Solid Waste Management Sector*. Boston: Massachusetts Institute of Technology, 2006.
[30] ADPC. *Flood Disaster Mitigation and River Rehabilitation by Marikina City,* Philippines. Safer Cities 22: Case studies on mitigating disasters in Asia and the Pacific. 2008.
[31] Dedehouanou, H., Coping with house waste management in Cotonou, *Environment and Urbanization*, **10(2),** p.191–208. 1998.

[32] Adelekan, IN Cities and climate change Global report on human settlements 2011, Washington/London Earthscan. 2011.
[33] Oresanya, O. *Integrated Waste Management Waste Collection & Transportation System* Paper presented at the C40 Summit Speaking Session 1st June, Curitiba, Brazil. 2011
[34] Tucci, C. E. M. Urban Flood Management. WMO and Capnet. 2007.
[35] Espinosa, M. *Policy and institutional aspects of water resources management in Nicaragua – The case of Managua.* Masters Thesis, Lund Universit: Lund, 2000.
[36] Ibarrarán, M. Climate's long-term impacts on Mexico's city urban infrastructure, Unpublished case study prepared for the Global Report on Human Settlements 2011, www.unhabitat.org/grhs/2011 cited in UN Habitat (2011) Cities and climate change Global report on human settlements 2011, Washington/London Earthscan, 2010.

Factors for the design of novel property-level flood resilient products: the Dado Wallboard

D. W. Beddoes[1,2] & C. A. Booth[1]
[1]*Department of Construction and Property, University of the West of England, Bristol, UK*
[2]*DrainAngel Ltd., Shropshire, UK*

Abstract

Community-level flood protection cannot guarantee property owners are free from flood risk, hence, it is the responsibility of property owners themselves to take appropriate action to ensure their properties are suitably protected by flood resistant and/or flood resilient measures. There are many flood products and avenues available for property owners to choose. This article introduces and describes the new *Dado Wallboard* property-level flood resilient product. It offers several practical advantages over existing internal building designs and fabrics, namely: (i) it is made from waterproof material and, as such, does not need to be replaced when a building has been flooded; (ii) it is mechanically fixed to the building wall substructure so that it remains in place when immersed in floodwater; and (iii) a movable dado rail and movable skirting board allow access into the cavity formed between the dado board and wall substructure and this allows both sides of a wall to be dried simultaneously. The product also offers socio-economic benefits, namely: (i) it reduces post-flood building reinstatement costs because fabric replacement is not necessary; and (ii) the wall drying procedure may reduce the overall time buildings are unoccupied.
Keywords: property adaptation, flood resistance and resilience, patent product.

1 Introduction

Flooding is one of the most dangerous and challenging issues facing humankind [1] and the catalogue of recent global disasters serve to exemplify this problem (e.g. Thailand (2011), Pakistan (2010), UK (2007) and the USA (2005), amongst many others). Effects of climate change on the built environment [2], coupled

www.witpress.com, ISSN 1743-3541 (on-line)
doi:10.2495/FRIAR120171

with increased societal pressure to further develop on floodplains, is overloading existing infrastructure and facilitating an increase in flood events. The impacts of these events can frequently encompass tragic loss of life and damage to both built and natural environments, plus massive insurance flood-repair claims [3, 4], considerable commercial businesses losses [5] and lengthy social disruption for property owners [6]. Therefore, governments and planning authorities are confronted with addressing the issues through suitable management strategies and appropriate legislation (e.g. Floods and Water Management Act (2010) in the UK) and through informed 'best practice' advice to property owners.

Where properties are known to be in a flood risk area, rather than ignore the situation and wait for a flood event to happen, there is an opportunity for property builders and/or owners to install flood resistance and/or flood resilience measures outside and/or inside buildings [7, 8]. Flood resistance measures attempt to halt floodwater entry into a property to prevent floodwater damaging its fabric [9, 10]. In contrast, flood resilience measures permit floodwater entry because the property is designed and constructed in a manner that floodwater impact is minimised, with no permanent damage caused, structural integrity maintained and subsequent drying and cleaning are easily facilitated [10].

This paper describes existing flood protection approaches used to adapt properties and outlines available property-level flood protection products, before introducing and detailing a novel flood resilient product (patent pending).

2 Existing flood protection measures

There are wide variations in the ways that people respond to probabilistic hazard forecasting, such as flooding, with some people making extensive preparation and others taking no action [11]. Unfortunately, there remains a very low take-up of property level flood protection. Surveys have shown that only a third of people who have experienced a flood take steps to protect their properties. This is mainly linked to concerns about: (i) perceived affordability of adaptation; (ii) anticipated impacts on property values; (iii) perceived costs and benefits of insurance; and (iv) social and emotional issues, such as anxiety, blame and regret, stigma and social identity [12].

It is recognised that flood protection products (Tables 1–3) need to be designed in such a way they not only protect properties from damage but also make the owner feel they are protected [12]. Flood protection choices for existing properties are between resistance and resilience measures, or a combination of both. External resistance measures are mostly acceptable to property owners but they do not always offer full protection; whereas, resilience measures are not a preferred choice for property owners, despite the benefits offered [13]. Minimising internal property damage is an obvious financial benefit but minimising the social disruption, by reducing the duration of the repair and reinstatement time, is an immense benefit for property owners because it clearly lessens the length of time spent away from the property. This is because flood resilience features offer the advantage to property owners that they can return and continue normal activities, after drying and cleaning, within a matter of days or weeks, as opposed to several months or sometimes more than a year after a flood event [8] or, in some cases, even two years [14]. Therefore, there is a need for sustained growth in the range of appealing and affordable flood resilient products for property owners to choose, without the attachment of emotional issues.

Table 1: Examples of permanent property-level flood resistant measures [15].

Permanent Flood Resistant Measures
1. Earth bund walls;
2. Periphery fences with sealed gates;
3. Raising building thresholds;
4. Storm porches to external doors;
5. External flood resistant doors;
6. Periscope or self-closing airbricks;
7. Anti-backflow valves on sewers;
8. Automatic activating external door guards;
9. Adding a water resistant render or extra skin of engineering bricks.

Table 2: Examples of temporary property-level flood resistant measures [15].

Temporary Flood Resistant Measures
1. Air-brick covers;
2. External door guards;
3. Flood skirts;
4. Mains unit flood alarms.

Table 3: Examples of property-level flood resilient measures [15].

Flood Resilient Measures
1. Internal tanking;
2. Concrete and tiled floors;
3. Raised electrical sockets and service meters;
4. Horizontal replacement plasterboard;
5. Resilient kitchen units (plastic, stainless steel, free standing, removable);
6. Raised kitchen appliances (ovens, fridges, freezers, washing machines);
7. Resilient internal walls (rendered, tiled, coated);
8. Plastic skirting board;
9. Plastic or acrylic internal doors or fitted with rising butt hinges;
10. Pump and sump systems.

3 Announcing a novel property-level flood-resilient product

The amount of damage caused to a building in a flood event is proportional to the depth of floodwater and the time the building is subjected to the floodwater [9]. In the case of a short duration flood event, external resistance measures such as ‘door guards’ can prevent some floodwater from entering a building but they cannot be safely used when floodwater depth exceeds 600mm because of the risk of structural damage [16]. When the differential height between building floor level and the height of the floodwater is over 600mm the hydrostatic water pressure can cause structural damage to the walls of the building and so the floodwater must be allowed to enter the building. Once the floodwater has entered the building the owners must wait for the floodwater to recede before beginning the clean-up and drying process to allow reoccupation.

Cleaning the building is relatively easy but drying out a building after floodwater has entered is a major problem [17]. Typically, the type of construction that needs drying out comprises sand/cement screeds, thick masonry walls, wooden skirting boards and gypsum plaster. Very often the gypsum plaster becomes damaged by the water and also contaminated by floodwater contents and so has to be completely removed. The wall substructure must then be dried out before re-plastering and then the new plaster has to dry out. Drying of wall construction takes a long time and whilst being dried the building has to remain unoccupied e.g. typical masonry construction dries at the rate of 1mm per day so that a 300mm thick wall takes several months to dry out. To overcome this, the present invention proposes a dado wallboard that forms a cavity against the wall substructure; an aperture at floor level allows access into the cavity for forced convective gas that constrained within the cavity passes directly across the surface of the wall substructure before exiting the cavity through another aperture part way up the wall.

Current drying methods comprise the removal of all the plaster on affected walls and then natural air drying whereby the wall substructure is dried with air circulated around the room to use natural drying potential. The process is very slow and weather dependent. It is a complex process involving consideration of room air temperature, relative humidity in the room, moisture content of wall substructure and temperature of the wall substructure being dried. An initial linear reduction of the wall substructure moisture content as a function of time the ‘constant drying rate period’ will give way to a ‘falling rate period’ and the time taken for drying increases accordingly. Directly applied heated gas convection will decrease the ‘constant drying rate period and lessen the ‘falling rate’.

Forced heated convective gas is blown into the cavity at floor level and passes directly across the surface of the wall substructure to increase the driving force for heat transfer and accelerate drying. Passing the heated convective gas intimately across the wall substructure also reduces air relative humidity that further increases the driving force for drying. Passing the heated convective gas intimately across the wall substructure will heat up the wall substructure and this higher temperature will speed up diffusion of water from the inside of the wall to the surface and so maintain a more continuous drying rate to accelerate the drying. The convective gas exits the cavity at the top of the dado board and as such is at the required height for leaving the room via an open window, so forcing the convective gas through the floor aperture and up through the cavity and then out via a window can be a rapid, efficient, whole room convective cycle for accelerated drying as opposed to the typical slow drying due to room air that is heavily moisture laden and re-circulated around the room.

The installation of the dado wallboard will look quite normal to the occupiers of the building. The dado wallboard can be finished with waterproof plaster and paint, the movable dado rail and skirting will both appear as normal. There will be no visual stigma associated with the dado board installation.

After the cavity and wall substructure has been sanitised via the apertures, and subsequently dried by forced heated convection, the room can be returned to use simply by replacement of dado rail and skirting boards, there is no need for re-plastering. The aperture formed between top of dado rail and normal wall covering can be set at any height required as long as it is above predicted floodwater heights. Typically when waterproof membranes are installed in wall construction and the floodwater is let into a building the floodwater manages to seep between the layers of construction. A film or layer of contaminated floodwater can become trapped between wall substructure and membrane and/or wall finishing materials and membrane. Often this contaminated water cannot be removed without first having to remove the wall finish (plaster) and/or membrane. The dado wallboard system renders all this remedial work unnecessary because removal of dado rail and skirting allows sanitising fluid to be introduced via the dado rail aperture into the cavity between the dado board and the wall substructure. The rear of the dado board and the surface of the wall substructure can both be accessed and sanitised with sanitising fluid flushing down the cavity and passing under the floor membrane to the sump/pump unit so that it can be pumped out of the building.

When floodwater levels are below 600mm the cavity formed between the wall substructure and the rear of the dado wallboard will collect any floodwater that passes through imperfections or faults within the wall at a higher level and ensure it falls downwards to the floor membrane for subsequent removal by the sump/pump unit. The cavity formed behind the dado wallboard can be used to house wall insulation. As part of the dado wallboard installation the building could be upgraded and fitted with wall insulation e.g. a closed cell insulation could be inserted behind the dado wallboard and removed for cleaning via the dado aperture in the event of a flood.

4 Detailed and diagrammatic descriptions of the product

The product, which is pending patent, is described and illustrated beneath. Figure 1 is a cross-sectional drawing to show the dado wallboard W. The drawing shows a building where the floodwater has reached a level that may cause structural damage and so the floodwater has been let into the rooms of the

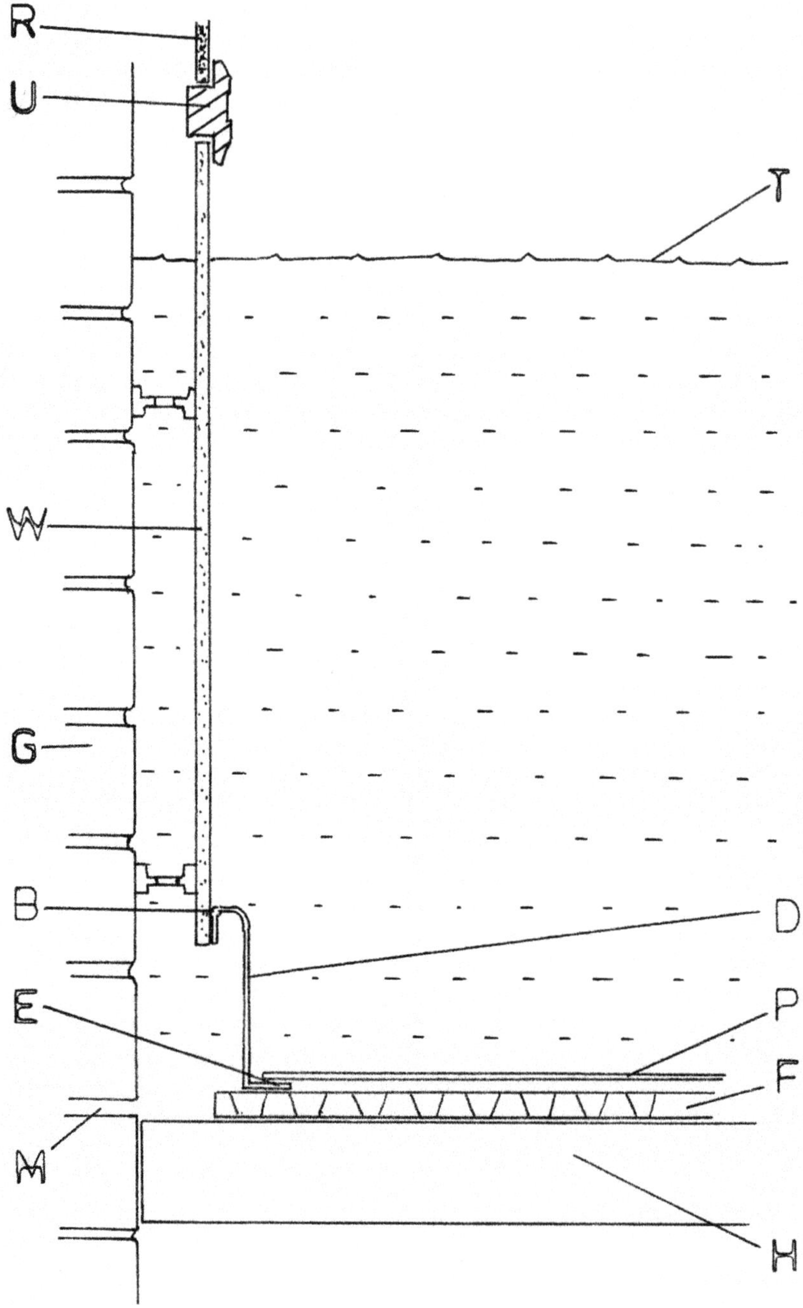

Figure 1: Cross-sectional drawing to show the dado wallboard fitted to a building and the floodwater has been allowed into the building.

building. The level of floodwater inside the building shown as T is now the same as the floodwater level on the outside of the building. The wall substructure shown as G is in this case masonry construction. The hole M is drilled through the inner skin of the wall to drain water from the centre of a solid wall or from a wall cavity. The waterproof dado wallboard W is mechanically fixed to the wall substructure. There is a removable or movable dado rail U fitted into or covering the aperture between the top of the waterproof dado wallboard and the conventional wallboard R. The aperture at the bottom of the dado wallboard is covered by the skirting board D that is fitted to the dado wallboard at B and the building floor at E. The floor covering P is laid over a floor membrane F that sits on floor construction H. The floodwater shown at level T fills the rooms within the building and the cavity formed between the surface of the wall substructure and the rear face of the waterproof dado board W.

Figure 2 is a cross-sectional drawing to illustrate the sanitising process of the wall structure after the floodwater has receded. The dado rail U is removed and aperture V allows access into the top of the cavity, similarly the skirting board is moved or removed to provide the aperture between dado wallboard and floor for access into the cavity between wall substructure and dado wallboard. Sanitising fluid C is introduced with applicator A into the cavity. Sanitising fluid is flushed into the cavity, as shown by arrows K, and completely cleans the cavity area including the surface of the wall substructure, the rear of the dado wallboard and any mechanical fixings in the cavity. The used sanitising fluid exits the cavity at L and passes under membrane F to a sump/pump unit for removal from the building. The dado rail and the skirting board can be movable or completely removable, all that matters is that access can be gained into the cavity by the apertures above and below the dado wallboard. The existence of the cavity and access into it makes it possible to sanitise and rapidly dry the structure. A waterproof membrane could also be incorporated into the wall construction but such a membrane must be of adequate profile and its location must be carefully considered so that it can be accessed for sanitising and drying i.e. the waterproof membrane if installed must not be able to harbour an inaccessible trapped layer of contaminated floodwater within the wall.

Figure 3 is a cross-sectional drawing to show the drying process of the wall structure after it has been sanitised. The forced hot convective gas S is entering the cavity via the lower aperture after the skirting has been moved. Some gas will enter via holes M to dry the wall and/or the wall cavity but the majority will travel up the cavity between the wall substructure and the rear of the dado wall board drying out the wall substructure and exiting through aperture V which is part way up the wall where the dado rail has been removed. The moisture laden gas leaving the cavity through aperture V is at a convenient height to then pass out of the building through an open window and set up a whole room convection cycle for accelerated drying. After drying out is complete the cavity can have any insulation replaced behind the dado wallboard by access through the apertures before the dado rail and skirting are refitted.

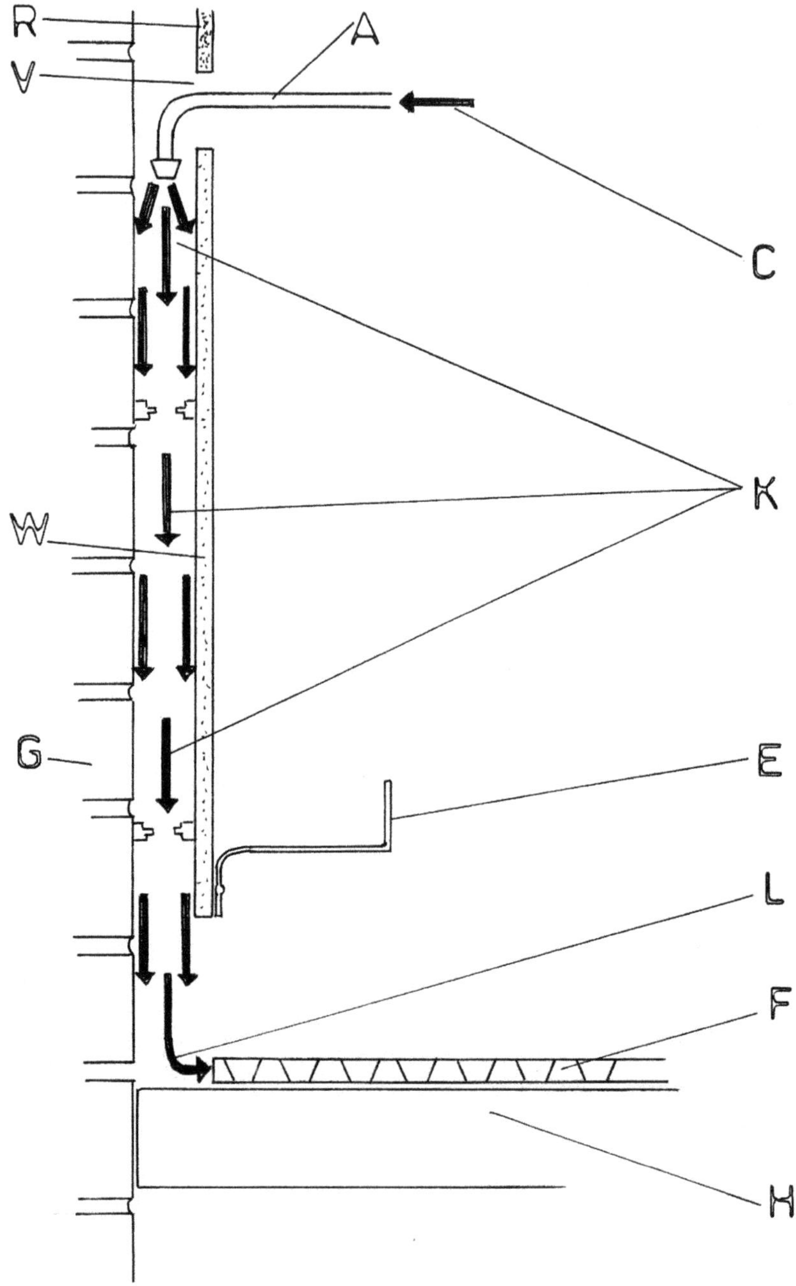

Figure 2: Cross-sectional drawing to illustrate the sanitising process of the wall structure after the floodwater has receded.

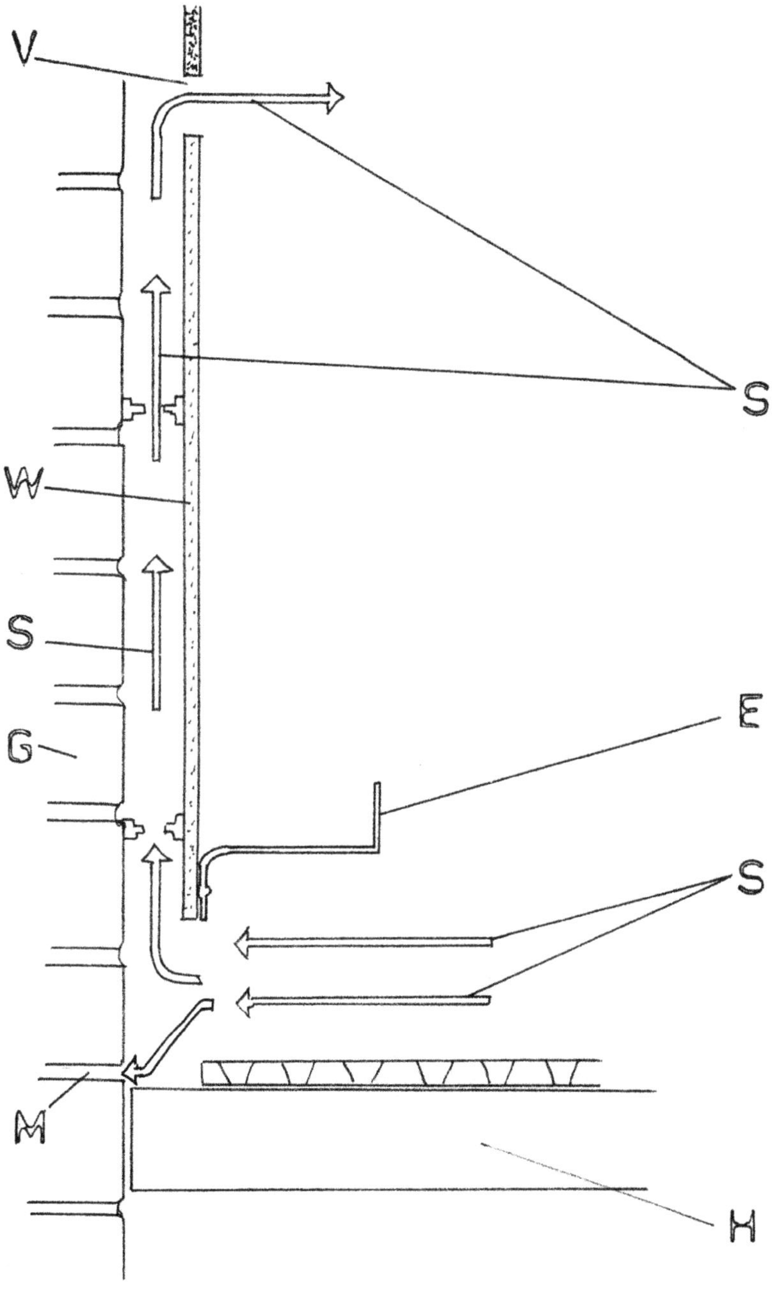

Figure 3: Cross-sectional drawing to show the drying process of the wall structure after it has been sanitised.

5 Comparable costs and benefits

Overriding determinants of product success are dependent upon cost and benefit. The costings [18] of three scenarios are described for the reinstatement of a standard ground floor room (4m x 4m) in a typical house that has been flooded to a depth of one metre. The first scenario (Table 4) is reinstatement to the original specification, the second scenario (Table 5) is reinstatement with a suite of flood resilient measures (as currently used by the flood repair industry) and the final scenario (Table 6) is reinstatement with dado wallboard flood protection.

Table 4: Scenario one – reinstatement of the room to the original specification – totalling £1937.60 [18].

Ref	Description	Quantity	Unit	Rate	Value £
1	Woodwork repairs and removals	16	m^2	7.13	114.08
2	Install new treated joists and floorboards	16	m^2	73.06	1168.96
3	Floor, wall and ceiling finishes	16	m^2	36.41	582.56
4	Sundries-electrical work	3	each	16.00	48.00
5	TV outlet	1	each	24.00	24.00

Table 5: Scenario two - reinstatement of the room with a suite of resilient measures – sub-totals at £2681.10 (plus the prime cost sum of £3000, totals £5681.10) [18].

Ref	Description	Quantity	Unit	Rate	Value £
1	Woodwork repairs and removal, plus sleeper walls and bake	16 1	m^2 sum	7.13 80.00	114.08 80.00
2	Excavation, earthwork and concrete work.	16	m^2	2.04	32.64
3	Filling with imported hardcore	24	m^2	15.74	377.76
4	Surface treatments	16	m^2	1.80	28.80
5	Expanded polystyrene board	16	m^2	8.90	142.40
6	Polythene dampproof membrane	16	m^2	1.39	22.24
7	In situ concrete floor slabs	16	m^2	211.80 per m^3	508.32
8	Cement screeds	16	m^2	30.06	480.96
9	Cement finishes - dubbing out	8	m^2	6.18	49.44
10	Cement finishes – including additive	16	m^2	40.02	640.32
11	Sundries-electrical work	3	m^2	50.02	150.06
12	TV outlets	1	each	54.08	54.08
13	Sump and pump unit, plus sewer backflow valve	1	sum	2000.00	2000.00
14	Aperture doorguard	1	sum	1000.00	1000.00

Table 6: Scenario three – reinstatement of the room with dado wallboard flood protection – sub-totals at £2553.74 (plus the prime cost sum of £3000 totals £5553.74) [18].

Ref	Description	Quantity	Unit	Rate	Value £
1	Woodwork repairs and removal, plus sleeper walls and bwk	16 1	m^2 sum	7.13 80.00	114.08 80.00
2	Excavation, earthwork and concreting	16	m^2	2.04	32.64
3	Filling with imported hardcore	24	m^2	15.74	377.76
4	Surface treatments	16	m^2	1.80	28.80
5	Expanded polystyrene board	16	m^2	8.90	142.40
6	Polythene dampproof membrane	16	m^2	1.39	22.24
7	In situ concrete floor slabs	16	m^2	211.80 per m^3	508.32
8	Floor, wall and ceiling finishes	16	m^2	6.87	109.92
9	Insulation for laminate floor panels	16	m^2	8.08	129.28
10	Dry-lining and partitions	16	m^2	43.50	696.00
11	Woodwork and UPVC second fixings.	16	m	6.76	108.16
12	Sundries-electrical work	3	m^2	50.02	150.06
13	TV outlets	1	each	54.08	54.08
14	Sump and pump unit, plus sewer backflow valve	1	sum	2000.00	2000.00
15	Aperture doorguard	1	sum	1000.00	1000.00

For simplicity, the following assumptions are inferred; the room has a suspended timber floor with tongue and groove boarding, the walls are `Gyproc` two coat wet plaster finish, electrics are three dual sockets with single TV point and any items that are common to all three scenarios (e.g. floor finishes, hacking off water damaged plaster, skirting boards and architraves) are not included in the cost comparisons. Also, the `prime cost sum` for installation of a sump/pump unit and sewer backflow valve has been entered (£2000) but a wide variation in costs exist [19]. Similarly, the `prime cost sum` for the aperture guard entered (£1000) can vary [20, 21].

The costings for the three scenarios show the least expensive option is to reinstate the room to its original specification; however, this cost must be met each time a flood event occurs. Furthermore, there will be sizeable costs associated with re-housing the flood victims (for several months) whilst all the works and extensive drying out periods are in progress [22]. Scenario two, reinstating the room with a suite of resilient measures (plus prime cost products), is more expensive but there are additional benefits because the insulated floor will not need replacement in the event of another flood and resilient walls together with a sump/pump afford some protection against low rates of water ingress. Scenario three, reinstating the room with dado wallboard flood protection costs slightly less than scenario two and offers many benefits for the homeowner. For instance, compared with scenario two, the homeowner will not be expected to sweep water ingress across the floor as it will pass under the floor membrane that is laid onto the concrete slab and then to the sump for removal.

There will be no rising damp in the walls as a result of water from below floors as a cavity is retained between wall and waterproof dado wall which can be ventilated via the skirting board and/or dado rail. In a flood event the system addresses residual risk as even when the differential height exceeds 600mm and water is allowed into the dwelling (achieved by letter box sited in door transom) there will be minimal damage if the floodwater stays below the designed dado rail height. The dado wallboard will allow quick and easy drying of the structure and the floor membrane with laminated floor finish are sacrificial at a cost of around two hundred and forty pounds. Dwellings with very porous external elements and/or Radon issues can incorporate perimeter floor drains to manage higher rates of water ingress at a cost of around two hundred pounds for the room as considered in this example.

6 Discussion

Floods are the most destructive natural disaster to affect humans [17], so it is important to remember the human misery and suffering caused by flood events. The majority of property owners who live in high-risk flood areas have not yet adopted any flood protection despite the high profile attention given to flood events by the media [23]. For instance, a leading insurance company, Norwich Union, recently surveyed 1500 UK flood victims and found most still believe there is nothing they can do to mitigate future flooding [24]. This is further supported by a survey conducted for the Department for Environment Food and Rural Affairs (DEFRA), which found, in areas of significant flood risk, only a small number of homeowners had taken any practical steps to limit potential flood damage [25]. Even those living in properties flooded several times before have taken only minimal action, often installing measures that are ineffective [25]. To address these issues, in 2011, the UK Government provided funding in excess of £2 million to protect 600 households based in 30 communities. Furthermore, in late 2011, the Environment Agency (EA) announced new grant aid of £1.3 million for a further 37 communities in England that are at risk of flooding, and this money is to be used to assist 500 homeowners wanting the installation of products to provide property level flood protection. More recently, the Flooding Minister (Richard Benyons MP) also promised at least another £2.1 billion will be spent on flood defence over the next four years [26].

In order to make the decision to install flood protection measures, the property owner must have both the desire to act and the ability to act [27].

Dhonau and Lamond [14] have detailed several case studies where property owners have eventually done something to minimise any future damage they may suffer; albeit, in one example, after suffering three floods in seven years and after being displaced for two years from their property. Nonetheless, by undertaking common flood resilience measures (Table 3) and creating some of their own (e.g. lightweight and easy to move furniture), they have accepted the problem is not going to disappear and they have now clearly had both the desire and ability to act. The case studies [14] highlight there are many ways to improve property-level flood resilience and the steps people are willing to take, and although the *Dado Wallboard* is just another flood resilience product, in addition to those already mentioned, if it is utilised by just one property owner and it reduces their damage and disruption, the authors will be duly satisfied.

7 Conclusions

The *Dado Wallboard* product demonstrates several practical advantages over those of existing internal building designs and fabrics, namely: (i) it is made from waterproof material and, as such, does not need to be replaced when a building has been flooded; (ii) it is mechanically fixed to the building wall substructure so that it remains in place when immersed in floodwater; and (iii) a movable dado rail and movable skirting board allow access into the cavity formed between the dado board and wall substructure and this allows both sides of a wall to be dried simultaneously. The product also offers socio-economic benefits, namely: (i) it reduces post-flood building reinstatement costs because fabric replacement is not necessary; and (ii) the wall drying procedure may reduce the overall time buildings are unoccupied.

References

[1] Lamond, J.E., Booth, C.A., Hammond, F. and Proverbs, D.G. (2011) *Flood Hazards: Impacts and Responses for the Built Environment*. CRC Press - Taylor and Francis Group, London.

[2] Booth, C.A., Hammond, F., Lamond, J.E. and Proverbs, D.G. (2012) *Solutions to Climate Change Challenges for the Built Environment*. Wiley-Blackwells, Oxford.

[3] Surminski, S. (2011) Flood insurance in the United Kingdom: the Association of British Insurers. In: Lamond, J.E, Booth, C.A., Hammond, F. and Proverbs, D.G. (Editors) *Flood Hazards: Impacts and Responses for the Built Environment*. CRC Press - Taylor and Francis Group, London, 89-97.

[4] Blanchard-Boehm, R.D., Berry, K.A. and Showalter, P.S. (2001) Should flood insurance be mandatory? Insights in the wake of the 1997 New Year's Day flood in Reno–Sparks, Nevada. *Applied Geography*, 21, 199-221.

[5] Ingirige, B. and Wedawatta, G. (2011) Impacts of flood hazards on small and medium sized companies. In: Lamond, J.E., Booth, C.A., Hammond, F.

and Proverbs, D.G. (Editors) *Flood Hazards: Impacts and Responses for the Built Environment*. CRC Press - Taylor and Francis Group, London, 269-278.

[6] Bell, D. (2011) Impacts of flooding in the built environment. In: Lamond, J.E., Booth, C.A., Hammond, F. and Proverbs, D.G. (Editors) *Flood Hazards: Impacts and Responses for the Built Environment*. CRC Press - Taylor and Francis Group, London, 37-51.

[7] Schelfaut, K., Pannemans, B., van der Craats I., Krywkow, J., Mysiak, J. and Cools, J. (2011) Bringing flood resilience into practice: the FREEMAN project. *Environmental Science and Policy*, 14, 825-833.

[8] Charlesworth, S.M. and Booth, C.A. (2012) Water resources issues and solutions for the built environment: too little versus too much. In: Booth, C.A., Hammond, F., Lamond, J.E. and Proverbs, D.G. (Editors) *Solutions to Climate Change Challenges for the Built Environment*. Wiley-Blackwells, Oxford, 237-250.

[9] Kelman, I. and Spence, R. (2004) An overview of flood actions on buildings. *Engineering Geology*, 73, 297-309.

[10] Bowker, P. (2007) Flood resistance and resilience solutions: an R&D scoping study. R&D Technical Report. DEFRA: London.

[11] Rose, C.B., Proverbs, D.G., Manktelow, K.I. and Booth, C.A. (2011) Improving community resilience: education, empowerment or encouragement? In: Lamond, J.E., Booth, C.A., Hammond, F. & Proverbs, D.G. (Editors) *Flood Hazards: Impacts and Responses for the Built Environment*. Taylor and Francis Group, London, 305-313.

[12] Harries, T. (2011) Why most "at risk" homeowners do not protect their homes from flooding. In: Lamond, J.E., Booth, C.A., Hammond, F. and Proverbs, D.G. (Editors) *Flood Hazards: Impacts and Responses for the Built Environment*. CRC Press - Taylor and Francis Group, London, 327-341.

[13] Tapsell, S. and Tunstall, S. (2008) "I wish I`d never heard of Banbury": the relationship between "place" and the health impacts from flooding. *Health and Place*, 14, 133-154.

[14] Dhonau, M. and Lamond, J.E. (2011) Property-level flood protection: case studies of successful schemes. In: Lamond, J.E., Booth, C.A., Hammond, F. & Proverbs, D.G. (Editors) *Flood Hazards: Impacts and Responses for the Built Environment*. CRC Press - Taylor and Francis Group, London, 291-303.

[15] Beddoes, D.W. and Booth, C.A. (2011) Property level flood resistance versus resilience measures: a novel approach. *International Journal of* Safety *and Security Engineering*, 1, 162-181.

[16] Johnson, M. (2011) Protection and performance of flooded buildings. In: Lamond, J.E., Booth, C.A., Hammond, F. and Proverbs, D.G. (Editors) *Flood Hazards: Impacts and Responses for the Built Environment*. CRC Press - Taylor and Francis Group, London, 259-267.

[17] Proverbs, D. and Soetanto, R. (2004*) Flood Damaged Property-A Guide to Repair*. Blackwell: Oxford.

[18] Langton, D. (2011) *Spon`s Architects` and Builders` Price Book 2011.* Abingdon: Spon Press.
[19] http://www.safeguardeurope.com/pdf_datasheets/sentry-twin-hla.pdf
[20] http://www.screwfix.com/p/door-guard-white-898-x-755mm/33363?_requestid=2045312
[21] http://www.floodguards.com/floodguards.aspx
[22] Garvin, S., Reid, J. and Scott, M. (2005) *Standards for the repair of buildings following flooding.* London: CIRIA.
[23] Harries, T. (2007) *Householder responses to flood risk; the consequences of the search for ontological security.* Unpublished PhD Thesis, Middlesex University, UK.
[24] www.aviva.co.uk
[25] Defra (2008) *Consultation on policy options for promoting property-level flood protection and resilience.* DEFRA: London.
[26] http://www.environment-agency.gov.uk/news/135003.aspx?page=2&month=11&year=2011&coverage=Anglian
[27] Proverbs, D.G. and Lamond, J.E. (2008) The barriers to resilient reinstatement of flood damaged homes. In: International Group for Research and Information on Post-Disaster Construction (2008) *4th International i-Rec Conference 2008 Building Resilience: achieving effective post-disaster reconstruction: April 30-May 2, 2008, Christchurch, New Zealand.*

Towards the development of a comprehensive systematic quantification of the costs and benefits of property level flood risk adaptation

R. Joseph[1,2], D. G. Proverbs[1], J. Lamond[1] & P. Wassell[2]
[1]*Faculty of Environment and Technology, University of the West of England, Bristol, UK*
[2]*Cunningham Lindsey, UK*

Abstract

Research in the UK, has shown that one of the reasons that people may not take action to guard against potential flood damage to their properties is that they lack first-hand information on the costs and benefits of available mitigation measures. From these perspective, fundamental issues of both universal and constraint uncertainties in property-level flood adaptation cost benefit analysis are discussed. Individuals who have direct knowledge of the potential flood risks that they are exposed to and subsequently have information on the costs and benefits of adapting their properties to flood risks, would be more likely to take action, and thus more inclined to undertake mitigation measures. The application of the concept of cost benefit analysis to flood mitigation measures at household levels has inherent uncertainties. A major exclusion in the past from flood mitigation cost benefit analyses has been the intangible impacts of flooding upon households, and this represents a form of systemic uncertainty. Research has shown that intangible impacts are both large and more important to affected households than are the tangible impacts; therefore quantification of the intangible impacts of flooding, for the purpose of developing a comprehensive cost benefit analysis model, is of a paramount importance in assessing the full impact of flooding on households, and hence currently represent a form of systematic uncertainty. The challenge of this research is how to quantify intangible impact/losses in the context of the cost benefit analysis approach.

WIT Transactions on Ecology and The Environment, Vol 159, © 2012 WIT Press
www.witpress.com, ISSN 1743-3541 (on-line)
doi:10.2495/FRIAR120181

A proposal for addressing this challenge is presented which could help homeowners living in flood risk areas in the UK to make an informed decision on investment in flood mitigating measures.
Keywords: cost-benefit analysis, choice modelling, contingent valuation, intangible impacts, property-level flood adaptation.

1 Introduction

It has been recognised, both at the European level and in the UK that there is a need for a shift from defensive action against flood risk to better management of flood risk. Flood risk management in the UK has, therefore, undergone a series of radical reviews (Institution of Civil Engineers *Learning to Live with Rivers*, 2001); DEFRA *Making Space for Water* (2005) and the Foresight Project *Future Flooding* (Evan *et al.* [1]) which have proposed less reliance on hard engineering solutions such as flood defence. Evans *et al* [1] stressed the need for a concept shift in which flood risk management relies less on Government intervention and more on an acceptance of individual responsibility. Concurring with this, the Environment Minister Richard Benyon said "*Government has a role to play in managing the overall risk to the nation, but we all, as individuals need to take greater responsibility for understanding our own flood risk and minimising it where we can*". In general, the most visible and obvious impact of floods upon households is the physical damage to contents and the fabric of the building [2] this is known as the tangible losses / impacts. These are damages, which can be easily specified in monetary terms [3].

This tangible damage is conventionally assessed in terms of such variables of flood characteristics such as depth and duration; land use patterns and the area's susceptibility to flood damage [4]. This conventional approach only takes into consideration the tangible damage, the consequences of which are that decisions are made with incomplete information in that intangible losses/impacts (for example impacts on health and general well being) are not considered. A further complication for UK householders in determining whether to invest in property level flood risk adaptation measures is that the bulk of reinstatement costs are currently borne by insurance companies for those homeowners who are fully insured [5, 6]. Insurance enables property owners to minimise the financial cost of damage from flooding but it cannot remove the trauma and misery of the experience.

Due to availability of insurance provision in the UK, some homeowners may have become complacent and have no reason to think of investing their limited resources in property-level flood adaptation measures, since it is believed that by paying their premium, they have successfully transferred the risk to the insurer. Although insurance can and does play a positive role, it may be more beneficial for property owners to reduce the effects of flooding as much as possible in the first place by protecting their homes from flood damage. This is because insurance provision only covers the tangible losses / impacts suffered by homeowners whereas the intangible looses / impacts are borne by homeowners as they are not insurable. Investment in property level flood adaptation measures have the potential to reduce some of the intangible losses / impacts which research has shown to be more devastating than the tangible losses / impacts [7]. As noted above, exclusion of intangible losses leads to a biased cost assessment of flood impacts/losses, and then an optimised allocation and design of damage reduction measures cannot be ensured. Hence, for a fully integrated assessment and management of flood risks it would be necessary to consider both the tangible and intangible impacts and their costs so as to incorporate these within the cost benefit analysis decision making tool.

Cost benefits analysis (CBA) is a widely accepted and often used tool for decision-making. CBA is an analytical tool based in welfare theory, which is conducted by aggregating the total costs and benefits of a project or policy over both space and time [8, 9]. To date there has been a dearth of research that investigates the possibility of applying the concept of cost benefit analysis to property level flood adaptation measures; this is due to difficulties involved in the process of monetising intangible impacts. One of the ways in which intangible costs could be included in the CBA decision making tool is to find a technique whereby intangible impacts/looses can be monetised so as to fit within the CBA decision making tool; this is one of the challenges of this research work.

It is against this backdrop, that a review and evaluation of the methods for quantifying the intangible impacts of flooding on households is hereby presented. Further, the context in which each of these methods can be used and their respective weaknesses are also explained. This aims at compiling and analysing the methods for the quantification of the intangible impacts/losses as well as to provide recommendations towards an effective application and subsequent incorporation of the intangible impacts in the CBA decision making tool at household property-level.

2 Methods of quantifying intangible impacts of flooding on households

Quantifying the cost of intangible impacts of flooding follows the main principles of welfare and environmental economics. The central themes in the field of welfare economics are to consider all categories of total economic value, including those generated by physical damages to properties and impacts on human health in decision-making. The conventional approach to flood damage mitigating measures takes into consideration only the tangible impacts; this is a determination that is considered to be insufficient for any sound decision-making by homeowners when considering investing in property-level flood adaptation measures. This in turn makes it important to quantify the intangible impact of flooding thereby increasing the accuracy of the assessment of the impact of flooding on households. To evaluate the role of the intangible damage component, there are economic valuation methods which are recognised as suitable to arrive at monetary estimates of non-monetary goods and therefore can be used to estimate the intangible impact of flooding on households. The most commonly used economic valuation methods are known as stated preference methods (SPM) and revealed preference methods (RPM) [9], both of which are considered below.

2.1 Stated preference method of quantifying intangible impact of flooding

Stated preference methods (SPM), also called direct valuation methods or inferential methods [9], have been developed to solve the problem of valuing intangible impacts/losses that have no obvious market value [10]. The SPMs use statistical techniques to infer the value of something that does not have an observable price [9]. They are survey or experiment-based approaches that elicit people's preference directly. There are two methods of stated preference techniques: contingent valuation method (CVM) and choice modelling method (CMM). Both methods use similarly structured questionnaires but differ in the way they define the non-market effect of concern. In addition to their ability to estimate user costs / values of any environmental impact on people, the most important feature of these survey-based methods is that they can be used to infer the value of intangible impacts/losses on households, thereby enabling the incorporation of the cost associated with intangible impacts within the CBA decision making tool.

2.1.1 Contingent valuation method (CVM)

Contingent valuation is a method of estimating the value that a person places on a good or service. Rather than inferring from observed behaviours in regular market places, the approach asks people to directly report their willingness to pay (WTP) to obtain a specified good or service, or willingness to accept (WTA) to give up a good or service. In the context of flooding; CVM depends upon a hypothetical situation or scenario whereby samples of flood plain residents are interviewed and individuals are asked to state their maximum willingness to pay (WTP) in order to reduce or eliminate the intangible impacts of flooding on their households. CVM is a social survey approach; therefore it is very important to pay special attention to the design and implementation of the survey questions and sample design [11].

The pre-requisites to using CVM in estimating the financial value of the intangible impacts of flooding on households are, individual survey of flooded residents, consultations with relevant experts, and pre-testing of the survey questionnaires [12]. Further, decisions need to be taken regarding how to conduct the interviews such as in-person, via mail or via telephone surveys. If the sample can be shown to be representative of the flooded residents population, the mean WTP that have been obtained from the sample can then be extrapolated across the population to obtain the combined WTP [11, 13].

The major strength of the CVM is that respondents are given the opportunity to state what they are willing to pay or what compensation they are willing to accept in order to reduce or eliminate the intangible impacts of flooding. Therefore if conducted appropriately the CVM will arrive at a measurement of equivalent financial value that is unambiguous and pertains only to the intangibles.

2.1.2 Weakness and bias of CVM

The possibility of strategic bias was a main objection to the use of CVM in quantifying intangible impacts among most economists. There are two forms of strategic behaviour, namely, free riding and over pledging [11]. Free riding would occur if an individual understates her true WTP for a public good on the expectation that others would pay enough for that good, and therefore, he/she need not have to pay. E.g, a rational response to the current flood risk management policy might be a tendency for respondents to understate the WTP in expectation that this will limit their future liability to contribute towards defence spending. Conversely, in evaluating individual property level measures which is not a public good, respondents may understate their WTP on the prospect that insurance company or government will pay for it. On the other hand, over pledging occurs when an individual assumes that their stated WTP value would influence the provision of good under question, provided that the stated WTP would not form any basis for the future pricing policy. For example, the situation in the past when decisions to build a flood defence from the public purse may have been based on the findings.

One of the weaknesses of using the contingent valuation method in the domain of property level measures is that questions are often problematic, as there is a lot of emotion involved in the answers to these questions because some of the questions may bring back the memory of the flood event(s). However a well designed questionnaire should be able to address this weakness. It is recognised that the WTP by individual is subject to their disposable income as there are income differential among the flood plain residents, this is another inherent bias towards the preferences of the better off. In order to address these drawbacks, a careful survey design is needed for income, age, and ethnicity.

2.1.3 Application of CVM in the UK

The most recent and noteworthy application of CVM to health related impact of flooding on households in the UK is *the Appraisal of Human Related Intangible Impacts of Flooding,* research which was jointly commissioned by Environment Agency and DEFRA (2004). The results demonstrate that flooding causes short-term physical effects and, more significantly, short and long-term psychological effects. More than 60% of flooded and at risk respondents expressed a willingness-to-pay (WTP) to avoid the health impacts associated with flooding. Based on the analysis of the survey, the authors recommended that the value of £200 per household per year be taken as representing the benefits of reduced health impacts as a consequence of a significant reduction in the risk of flooding.

While this can be seen as a move in a positive direction with regards to the quantification of intangible flood impacts on households, the value of £200 appears to be unrealistic if this is compared to the value of tangible impacts. For example, Green and Penning-Rowsell [14] carried out experimental work by using a bootstrapping procedure to relate known tangible losses to intangible losses by deriving equivalent values. This method was based upon an extensive interview survey with flood victims. In conducting the interview, affected households were asked about the nature of the flood which they have experienced, what effects it had and how much it cost to repair or replace the damage, most importantly the respondent were asked if they recovered all their costs from insurances and other sources. The study found that the monetary equivalents of the intangible impacts far exceeded the tangible impacts. Further, Adamson [15] indicated in his report that the intangible impacts are currently set as equivalent to the residential property damage in the evaluation of potential flood damage in the Republic of Ireland.

2.1.4 Choice modelling method (CMM)

Choice modelling is concerned with the individual attributes of, say, a flood and estimates WTP for these individual attributes. This method assumes that each respondent has a perfect discrimination capability, whereas the analyst has incomplete information and must therefore take account of uncertainty [10]. The respondents are presented with different bundles of flood impacts which can be reduced by implementation of property level flood alleviation measures at a certain price among which they are asked to make a choice. One of the characteristics of choice modelling methods is a price, and this enables the analyst to infer willingness to pay without asking directly what it is [16].

Similar to contingent valuation method, the cost modelling method can estimate financial values for any environmental resource, and can be used to estimate tangible as well as intangible values. CMM however, enables estimation not only of the value of the environmental resource as a whole, but also of the inherent value of its attributes, their implied ranking and the value of changing more than one attribute at once [17, 18]. There are a number of variants to choice modelling which differ in the way they present the respondent with the choice task. Only two of these comply with the requirements of economic theory: choice experiments and contingent ranking. Choice experiments require the respondents to choose their most preferred scenario. Contingent ranking requires the respondents to rank the given scenarios according to their preference.

2.1.5 Application of CMM in the UK

In the study of *the Appraisal of Human Related Intangible Impacts of Flooding in the* UK which was jointly commissioned by the Environment Agency and DEFRA, the choice modelling method was used to estimate how much people are willing to pay for a marginal improvement in each attribute of a flood and to estimate how much people are willing to pay to avoid a given flood so long as this flood is within the range of scenarios presented to the respondent. One of the attributes was always price or cost so that WTP was inferred through econometric analysis of trade offs people make between other attributes and cost.

2.1.6 Choice modelling versus contingent valuation methods

CMM has become more popular due to several advantages over CVM. These include; the ease of estimating values of single attributes of a flood impact on households; avoidance of part-whole bias problem since different levels of the impacts can easily be built into the experimental design; respondents are more familiar with making choices rather than generating spontaneous valuations. Moreover, CMM can solve some of the biases that are present in CVM; the strategic bias which is associated with CVM is minimised in the CMM since the prices of the different intangible flood impacts are already defined in the choice sets. However, the choice approach does limit the valuation to predefined options and does not allow for the respondent to state a zero value for the resource even if they do not value it, however, in order to eliminate the limitation, *no change no expense* can be an option within the choice set.

2.2 Revealed preference method of quantifying intangible impact of flooding

Revealed preference methods, also known as indirect valuation methods, look for related or substitute markets in which the environmental good is implicitly traded. Information derived from observed behaviour in the substitute markets is used to estimate willingness to pay (WTP), which represents individual's valuation of, or the benefits derived from, the investment in flood protection measures [9]. The revealed preference methods have the advantage of producing estimates of the value for a particular good from actual market behaviour; however, care must be taken in extrapolation to market conditions which have not been observed in practice. The two most popular such methods prevalent in the environmental economics literature are the hedonic pricing and the travel cost methods [3, 5, 9, 10]. In the context of quantifying the intangible impacts / looses of flooding on household, only the hedonic pricing methods will be considered, because travel cost methods are based on observed behaviour, it is used to estimate use values only.

2.2.1 Hedonic pricing method

The hedonic pricing method is intended to determine the influence of the area's characteristics in question, such as a sea view, on the market price of the land or more usually, houses. At the most basic level, a hedonic property value model is based on the premise that it is possible to establish a statistical relationship between a home's sale price and its characteristic. The assumption of the technique of hedonic price is based on the assumption that goods are valued on the basis of a utility provided by their own characteristics. They can be revealed by an observation of property's market prices in relation with the amount observed by attributed characteristics of each property. Macdonald *et al* [19] carried out a study linking flood risk insurance and the land market to the hedonic price method. They evaluated the consumers' willingness to pay after a reduction of floods occurrence probability. In their models, the hedonic prices are the implicit prices of attributes, such as projects to reduce the flood risk and insurance.

2.2.2 Application of hedonic pricing methods

In the studying of the effect of floodplain disclosure in California, Troy and Romm [20] built a multi-layered hedonic regression model containing 27 to 30 significant independent variables including flood status variables, property characteristics such as number of bedrooms, locational characteristics such as coastal location and socio-economic variables such as median household income.

Hedonic method was applied in Schleswig Holstein (Germany) to derive estimates of the value people attach to different coastal attributes that are at stake if flood events increase as a result of climate change [21]. In Glasgow studying the effects of noise pollution on house value, Day [22] included 31 significant variables including property characteristics, locational characteristics and amenity variables such as views from the property. Lamond *et al* [23], recognised that it is apparent that to explain adequately the variation in house price between one property and another a great variety of causal variables are needed, and equally that large numbers of observations are necessary to estimate such complex models with any degree of confidence.

There are inherent problems in using the revealed preference methods in quantification of intangibles; the first problem is that the method will measure the flood risk effect as a whole and it is practically impossible to identify what is being taken into account. In order to value intangible in isolation, there will be a need to have some households which experienced intangible losses but no tangible ones and the properties which such household occupied would need to be known, identifying such households may posses problem. The second problem is that of information asymmetry which, arguably, is one of the reasons that no impact is measurable in the long term UK property market.

3 Incorporation of monetary value of intangible impacts/losses in CBA decision making tool at property level

The application of CBA to real world problems is not always easy. In quantifying the change in an individual's utility as a result of a flood, two strategies can be used. One, a global or holistic assessment of the change and two, a piecemeal approach of identifying, measuring and quantifying every impact of the flood upon individual or households [14]. The second approach is normally adopted in cost benefit analysis (CBA) approach on flood mitigation measures in the UK.

The aim of CBA is to determine whether a project is desirable from the point of view of social welfare, by means of the algebraic sum of the time-discounted financial costs and benefits of the project. The economic efficiency of the project with long-run-accruing costs and benefits is often assessed using the net present value (NPV) rule. In the context of applying the CBA to property-level flood adaptation measure, there is a requirement that all impacts are taken into account. However, in the analysis of the costs and benefits of flood alleviation measures in practice, the analysis is often narrowed to the consideration of tangible monetary impacts. For example Joseph *et al* [24] established the cost effectiveness of resilient reinstatement, the result of the analysis indicates that resilient reinstatement is cost beneficial to floodplain resident with 5-25 years flood frequency, this conclusion was reached by considering tangible impacts of flooding. It can be inferred that, if intangible impact was included within the analysis, there is the possibility that the adoption of resilient reinstatement would be economical for properties within a floodplain area of up to 50 years flood frequency. The application of CBA to property-level measures means that both costs and benefits need to be broken down in a very detailed way and this means that revealed preference technique is not appropriate because it cannot be broken down.

The practicality of quantification of all costs and benefits is still posing problems in the context of flood protection at an individual property level. Therefore any step towards making the quantification possible makes the comparison and decision more robust. Yet, there is no generally accepted framework available when considering application of the concept of CBA at property-level, where the relevant pieces of information are put together, and there are different ways of monetising the non-monetary impacts. To quantify the intangible impact of flooding on households in pursuit of a comprehensive CBA process for adopting property-level flood adaptation is to use stated preference method of environmental valuation as discussed in sections 2 and 3 above.

3.1 The use of a CBA tool at household level

Presently, there is no evidence to suggest that floodplain residents will willingly embrace responsibility for protecting their properties against flood risk because it is not in their financial interest to do so [25]. However, despite their limitations and criticisms, a cost benefit analysis tool could provide detailed property specific information to the homeowner when considering adopting property-level flood adaptation measures. The CBA tool could be used by homeowners during post flood recovery periods when reinstatement work is about to be carried out or during a planned renovation work. Further there is a potential for the CBA tool to be used in non-flood situations and for new build. There is consensus within the literature that implementing resilient or resistance measures during planned renovation work is cost effective [6, 26].

The development of a CBA tool for flood mitigation measures could be a useful decision-making tool for homeowners considering to adapt their property to flood risk. For some homeowners, cost will be an absolute barrier. The margin between their basic living costs and their income will not allow them to invest monies on property-level measures. However, with a well structured CBA tool which incorporates all flood impacts on households, there is a possibility of lenders recognising the cost and benefits of the adoption of property-level flood adaptation measures, thereby being willing to offer loans to cover the cost of property-level measures. For homeowners who can afford to invest in property-level measures, the CBA tool could help decide which adaptation solution is most effective on a CBA basis. The CBA tool which incorporates both impacts of flooding (tangible and intangible) can be extended to commercial properties, while this review is based on UK experience, the result of the review can have international application, while the building characteristics may be different from country to country, most of the intangible impacts are expected to have similarity from region to region, thereby making the international application of the full CBA tool possible.

4 Conclusions

So far, the conventional approach has not paid much attention to the possibility of quantifying intangible impacts of flooding on households with the intention of incorporating this in the CBA tool for decision making on investment in property–level flood measures. The results of this theoretical review will lead to improved methods for evaluating the intangible impacts of flood on households. The incorporation of the full impact of flooding in the CBA decision making tool at property-level has the potential to lead to a greater understanding of the full costs and benefits of property-level flood mitigation measures by different stakeholders, (i.e. homeowners, lenders and possibly insurance companies). Thereby helping homeowners in decision making on the appropriate adaptation solution which is most effective to their property and situations on a CBA basis. Two methods of valuations of intangible impact of flooding on households were identified in this review, stated preference method (SPM) and revealed preference methods (RPM). The stated preference methods (SPM) are seen to be more suitable than the revealed preference methods (RPM), for monetising intangible impacts of flooding on households. The SPM has their weaknesses and limitations. However the use of SPM valuation methods in eliciting the cost of intangible impacts on households enhances the ability to incorporate common monetary unit within a sound theoretical basis of CBA.

Finally, it has been identified theoretically the importance and methods of monetising intangible impacts of flooding on households and the benefits of incorporating this in the CBA decision making tool at property-level. Further research which will apply the valuation method identified in this review to a real life flood event and subsequent incorporating the monetised intangible impacts into the CBA decision making tool at household property-level is proposed.

References

[1] Evans, E. P., Ashley, R., Hall, J., Penning-Rowsell, E., Sayers, P., Thorne, C. R. and Watkinson, A. *Foresight. Future Flooding,* **1 and 2** London. 2004.

[2] Joseph, R., Proverbs, D., Lamond, J. and Wassell, P. A critical synthesis of the tangible impacts of flooding on households. *27th ARCOM annual conference 5-7th September 2011.* University of the West of England, Bristol, UK. 2011(c)

[3] Markantonis, V. and Meyer, V. Valuating the intangible effects of natural hazards: a Review and evaluation of the cost-assessment methods. *European Society for Ecological Economics conference.* Istanbul 14-17 June. 2011.

[4] Lekuthai, A. and Vongvisessomjai, S. Intangible Flood Damage Quantification. *Water Resources Management*, **15**(5), pp. 343-362. 2001.

[5] Lamond, J. E. *The impact of flooding on the value of residential property in the UK.* Doctor of Philosophy (PhD). Thesis, University of Wolverhampton. 2008.

[6] Wassell, P., Ayton-Robinson, R., Robinson, D., Joseph, R., Hack, K., Butler, D., Salkeld, I. and Twomey, J. *Resilient Reinstatement: The costs of flood resilient reinstatement of domestic properties.* Association of British Insurers. 2009.

[7] Joseph, R., Proverbs, D., Lamond, J. and Wassell, P. A critical synthesis of the intangible impacts of flooding on households. *International conference in building resilience: Interdisciplinary approaches to disaster risk reduction and the development of sustainable communities.* Sri Lanka. 2011.

[8] Hanley, N. and Spash, C. *Cost benefit analysis and the environment.* England: Edward Elgar Publishing. 1995.

[9] Penning-Rowsell, E., Johnson, C., Tunstall, S., Tapsell, S., Morris, J., Chatterton, J. and Green, C. *The benefits of flood and coastal risk management: A manual of assessment techniques.* Flood Hazard Research Centre: Middlesex University Press. 2005.

[10] Birol, E., Karousakis, K. and Koundouri, P. Using economic valuation techniques to inform water resources management: A survey and critical appraisal of available techniques and an application. *Science of The Total Environment*, **365**(1-3), pp. 105-122. 2006.

[11] Mitchell, R. C. and Carson, R. T. *Using surveys to value public goods: the contingent valuation method.* Baltimore: Johns Hopkins University Press. 1989.

[12] Hanemann, M. Valuing the environment through contingent valuation. *Journal of economic perspectives*, **8**(4), pp. 19-43. 1994.

[13] Environment Agency and DEFRA *The Appraisal of Human Related Intangible Impacts of Flooding.* Joint DEFRA/EA R&D technical report FD2005/TR, DEFRA/EA. 2004.

[14] Green, C. H. and Penning-Rowsell, E. C. Flooding and the Quantification of 'Intangibles'. *Water and Environment Journal*, **3**(1), pp. 27-30. 1989.

[15] Adamson, M. Floods: causes, management and relief. *The Chartered Institution of Water and Environment Management.* Ireland Office of Public Works. 2003.

[16] Brouwer, R. and Pearce, D. W. *Cost-Benefit Analysis and Water resources Management.* Cheltenham: Edward Elgar. 2005.

[17] Hanley, N., Wright, R. and Adamowicz, V. Using Choice Experiments to Value the Environment. *Environmental and Resource Economics*, **11**(3), pp. 413-428. 1998.

[18] Bateman, I., Carson, R., Day, B., Hanemann, W. M., Hanley, N. and Hett, T. *Guidelines for the use of stated preference techniques for the valuation of preferences for non-market goods.* Cheltenham: Edward Elgar. 2003.

[19] Macdonald, D. N., Murdoch, J. C. and White, H. L. Uncertain hazards, insurance, and consumer choice: evidence from housing markets. *Land Economics*, **63**(4), pp. 361-371. 1987.

[20] Troy, A. and Romm, J. Assessing the price effects of flood hazard disclosure under the California natural hazard disclosure law. *Journal of Environmental Planning and Management*, **47**(1). 2004.

[21] Hamilton, M. J. Coastal landscape and the hedonic price of accommodation. *Ecological Economics*, (62), pp. 594-602. 2007.
[22] Day, B. Submarket identification in property markets: a hedonic housing price model for Glasgow, working paper. EDM 03-09, CSERGE, University of East Anglia. 2003.
[23] Lamond, J., Proverbs, D. and Antwi, A. Measuring the impact of flooding on UK house prices A new framework for small sample problems. *Property Management*, **25**(4), pp. 344-359. 2007.
[24] Joseph, R., Proverbs, D., Lamond, J. and Wassell, P. An analysis of the costs of resilient reinstatement of flood affected properties: A case study of the 2009 flood event in Cockermouth. *Structural Survey*, **9**(4), pp. 279-293. 2011.
[25] Lamond, J. E. Financial implications of flooding and the risk of flooding on households. *in* Lamond, J., Booth, C., Hammond, F. and Proverbs, D. (eds.) *Flood Hazards: Impacts and Responses for the Built Environment.* London: Taylor and Francis Group, pp. 317-326. 2012.
[26] ABI Assessment of the cost and effect on Future Claims of Installing Flood Damage Resistant Measures. Association of British Insurers: London. 2003.

"Three times is enemy action" – flood experiences and flood perceptions

C. B. Rose[1], D. G. Proverbs[1], C. A. Booth[1] & K. I. Manktelow[2]
[1]*University of the West of England, UK*
[2]*University of Wolverhampton, UK*

Abstract

In the context of anticipated climate change, particularly the frequency and intensity of rainfall events likely to affect the UK in future, it is becoming increasingly important to understand the ways in which people perceive, and therefore respond to, natural hazards such as flood risk. Behavioural models, incorporating key predictive factors, are sought by the policy-making bodies, in order to develop effective interventions for improving community resilience to extreme weather events. An examination of factors relevant to flood risk perception is presented, including the degree of exposure to the threat; concepts such as 'controllability' and 'dread' associated with different hazards; and the wider issues of perceived responsibility, blame-shifting and ontological security. The complex interactions between the factors, together with some UK-specific issues, pose challenges for the development and testing of any model seeking to explain and predict the behaviour of people in the UK exposed to flood risks. Some innovative techniques, capable of modifying aspects of the perception of risk, which can be incorporated into community engagement initiatives, are also discussed.
Keywords: risk perception, flood preparedness, natural disasters, ontological security, community engagement.

1 Introduction

The UK, amongst other nations, has a problem in respect of the population known to be at flood risk:

"... many householders simply do not think that flooding will affect them, even though they are well aware that they live in the threatened area" [1].

WIT Transactions on Ecology and The Environment, Vol 159, © 2012 WIT Press
www.witpress.com, ISSN 1743-3541 (on-line)
doi:10.2495/FRIAR120191

This essentially represents a mismatch between the perceptions of the general public and the evidence-based risk perceptions of officials, professionals and technical experts. In the context of nuclear power plants, this perception gap has been attributed by the industry's proponents to 'public ignorance and irrationality' 2] but this is an overly simplistic view. The oft-proposed remedies, of education and awareness-raising, fail to address the complexities of the human thought processes, emotions and other factors that contribute to our perceptions of risk.

More recently, policy-makers have sought a deeper understanding of how misperceptions can be corrected and behavioural changes brought about [3] but no 'magic bullet' exists, as the realities of human behaviour, and techniques for modifying it, are complex issues. In the context of flood warning responses, for example, the public's assessment of risks has been found to combine the probability estimates provided by a warning authority with a wide variety of other information: this can include beliefs, recent experiences and perceptions of the credibility of authorities, amongst other factors [4].

It has also been observed that people have a marked predilection for viewing themselves as personally immune to hazards, as illustrated by the following finding:

"The great majority of individuals believe themselves to be better than average drivers" [2].

This perception is, in statistical terms, an impossibility but there is an experiential explanation: behaviours such as contravening speed limits or following too close behind another vehicle will lead to an accident (or other adverse consequences) on *relatively few occasions*, thereby creating a body of personal experience that supports the individual's belief in their competence [2].

To explain this, some researchers have adopted a 'mental models' approach to explain risk-related behaviours; this posits that people develop conceptual structures to make sense of their world, which may well be inaccurate, or include misunderstandings [5]. The models adopted by a member of the public are, therefore, likely to differ from those of a professional working in the relevant field. Furthermore, such mental models are not limited to an individual's perceptions, but may also be shared amongst a community, leading to what might be termed 'myths and legends' pertaining to hazards: for example, in a US study, local residents asserted that a flood event had been 'man-made', because of a commonly held, but erroneous, belief that an upstream hydro-electric dam was a flood-control structure and that this had 'failed' [6]· Such misunderstandings are not limited to the general public: a publication for the Geological Society of London includes the description: 'Clywedog Dam at Llanidloes, a flood control dam on the River Severn...'[7]; the structure in question is, in fact, a water supply reservoir with extremely limited, localised flood alleviation capability [8].

Attempts to conceptually model human behaviour in risk contexts, therefore, face a dual challenge: not only are there multiple factors involved, but also there is interaction between factors. A generic framework for social psychological analysis, designed to aid experimental work in this field, has been developed by

Breakwell [8]: this offers ten broad factor-groupings ranging from physical/environmental contexts to ideology/social representations. Of particular interest in the present context, however, are two groups which have been implicated in the perception of flood risk: the first encompasses factors arising from thoughts or feelings; the second, termed 'past action', subsumes both prior experience and previous behaviours These factor groups contribute to some of the phenomena that have been observed in the context of natural hazards including flooding: the misattribution of blame; the concept of 'dread risk'; and the interactions of flood experiences and preparedness. The following sections examine these issues in depth, with the intention of synthesising the extant literature on preparedness to hazards, in order to develop a framework specific to flood risk preparedness behaviours in UK citizens.

2 Thoughts and feelings: blame and controllability

' ... flood victims blame engineers for holding back water to protect London, just as Point Marion residents said about Pittsburgh ...'[6].

The blame discourse is widely recognised as an issue amongst researchers in the field of flooding: based upon the mistaken belief that flooding is an entirely controllable phenomenon, people will invoke human incompetence (or malice) as being the causes of a flood event [10] yet, by contrast, seismic or volcanic hazards are perceived as natural and uncontrollable phenomena. If it is believed that some preventative actions have not been carried out, anger and frustration have been noted as common responses in affected communities [6, 11]. By seeking to shift blame onto others (typically local or national authorities) denial of personal responsibility can be maintained, not only in the context of natural hazards [12] but also in issues such as engagement with initiatives to address climate change [13].

Additional emotionally-laden nuances apply to hazards that directly affect the home, one such being the concept of ontological security [14]. Otherwise known as 'the security of being', this is posited to be rooted in the unconscious mind and closely linked to an individual's self-identity. The outside world is here perceived as being threatening and uncontrollable in nature, but the home is, by contrast, a place of safety. Flooding would, in this case, constitute a violation of 'feeling safe' stemming from an uncontrollable external force; denial that flooding is a possibility could, therefore, be seen as a means whereby this sense of security can be upheld [10].

As well as the degree of controllability associated with a given threat, factors such as the level of familiarity with a hazard have also been found to contribute to the way in which different risks are perceived: known hazards, for example, are judged to be of less concern than unknown, or novel, threats [15].

3 Thoughts and feelings: the unknown and the dreaded

In a major study on the perception of 30 different hazard types, ranging from nuclear power to home appliances, four contrasting groups of subjects were asked to rate all the activities on a list of characteristics hypothesised to be of relevance [16]. The results indicated that these could be consistently grouped into just two underlying factors as follows:

1. The 'dread risk' factor comprised degree of controllability, dread, catastrophic potential, fatal consequences, and the inequitable distribution of risks and benefits

2. The 'unknown risk' factor was characterized as unknown, new, delayed manifestation of harm and unobservable consequences.

The relative positions of four hazards on these two factor scales are illustrated in Figure 1. In this representation, nuclear power scores high on both factor groups and so appears in the upper right quadrant; crime, though seen as familiar is also relatively uncontrollable and potentially fatal and so falls in the lower right quadrant. Neither aspirin nor bicycles are perceived as a dreaded risk, but the risks arising from the drug are felt to be less well known than those associated with the means of transport.

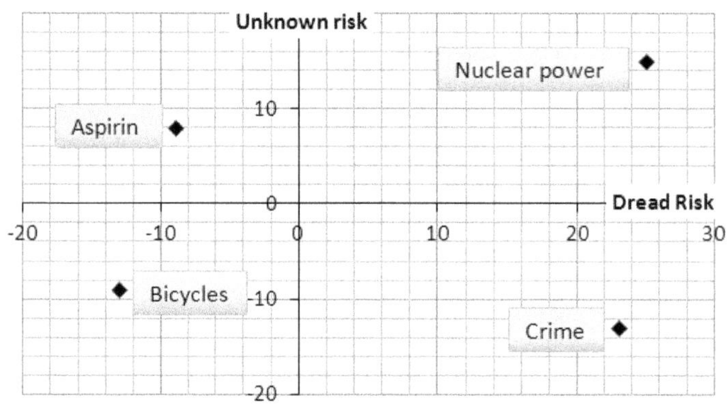

Figure 1: Location of hazards within a two factor space (adapted from Slovic *et al.* [1]).

The two contrasting approaches were compared in a meta-analysis of studies on the relevance of trust to the perception of risk [21]; neither approach was strongly supported, the conclusion being that the relationship hinges on contextual factors. The Psychometric Paradigm has been more widely utilised by other researchers, however, with the 'dread risk' factor group being found to be of particular salience to natural hazards perception. Examples include an increased likelihood of preparing for hurricanes when the individual engages with institutions that communicated urgency and dread [22] and evidence of strong correlations between dread levels and previous flood experiences [23].

The thoughts and feelings of individuals, therefore, constitute relevant factors in their perception of risk; these are derived, at least in part, from both direct and vicarious experiences of hazards. The effects of personal hazard experience on protective behaviour have been the subject of many studies, and this topic will now be explored.

4 Past action: hazard experience and preparedness

'Experience leads people to think about the risk more often, and with greater clarity' [24].

Past exposure to a given hazard might seem, intuitively, to provide an obvious driver for future mitigation actions, yet research shows this to be an overly simplistic expectation. Experience with volcanic hazards, for example, did not necessarily motivate individuals to make preparations for future crises of the same nature [25], whilst a study of evacuation behaviour during Hurricane Katrina found that the influence of friends and family members was more important than prior experience [26]. Having survived a major storm (hurricane or cyclone, depending upon the geographical location) was, however, a contributory factor in households having prepared evacuation plans [27], emergency supply kits (such as bottled water and portable radios) [27, 28] and practices such as deploying shutters on windows and doors [22].

This is a somewhat mixed picture could suggest that hazard experience *per se* does not offer a consistent motivating factor; it has been suggested that, in part, the wide range of activities that can be subsumed under 'preparedness' have militated against the emergence of a coherent picture. A study that separated these activities into two groups, those linked to 'survival' (first aid kits, water and the like) and 'damage mitigation' (such as securing tall furniture to walls in seismic risk areas) found that the first group of actions were more commonly adopted than the second [29]. Furthermore, hazard exposure can arise from two sources: direct interaction with the threat, or indirect (vicarious) experience from media reports, or hearing vivid descriptions of others' experiences [30, 31]. It is possible that the small minority of people reported as having taken anticipatory actions in the absence of any personal experience may, therefore, have been prompted by vicarious experiences. The latter is a factor identified as contributing to the cognitive processing of self-efficacy (perception of the competency to act) [32] which, in itself, has been implicated in some explanatory models of preparedness [33, 34]. This would suggest that work towards a UK-specific preparedness framework could usefully include questions on both types of experience to tease out any such effects, as well as clearly distinguishing between different preparedness categories.

5 Past action: 'three times is enemy action'

The frequency of exposure has been found to be of relevance as a predictor of protective behaviour in the UK: when the number of experiences of household flooding rises above two or three, the frequency of mitigation behaviours is found to increase greatly [35]. Although one incident of flooding may be dismissed by householders in many cultures as the workings of 'chance' or fate, repeated floods have an additional complexity for UK householders, owing to the practice of flood insurance cover being 'bundled in' with building and contents cover [36]. For a first claim, the resultant repair and renovation costs are likely to be met, but insurance companies may well raise both the premium and the excess in respect of any future flood claims [37]; the implication is clear, once may be an accident, but a second claim is not simply 'coincidence' from the insurers' standpoint. If further flood incidents occur, even those who have been vociferous in denying the existence of flood-risk in the past may come to the acceptance that 'enemy action' is indeed taking place, and investment in flood resistance or resilience measures is required. In models of flood hazard from other cultures, flood insurance is more usually purchased as a separate policy; future research on UK-specific flood-risk populations would, therefore, benefit from incorporating this aspect within the conceptual model adopted.

As we have seen, the overall picture is indeed complex, yet policy-makers must still attempt to manage down the risks posed by flooding, and to achieve this by demonstrating 'greater empathy' [38]: one approach that is being increasingly adopted is that of direct engagement with the people at risk, in order to improve resilience.

6 Community engagement programmes

A community consists of a group of individuals who are connected to one another in a range of ways, including geographical, cultural or other common interests. Where communities undergo experiences that improve their perceived control over their environment, the sense of 'belonging' can contribute to the adaptive capacity of individuals within that group [39]. This represents an interaction between perceived self-efficacy (of the individual) and the collective efficacy of the community as a whole. By entering into activities that empower communities, such as participation in decision-making, and supporting community-led initiatives, the policy-making institutions create an opportunity to address the 'thoughts and feelings' issues discussed above such that problems can be articulated and misconceptions corrected. This approach, therefore, far exceeds the simple provision of information: by understanding the needs of communities and providing meaningful assistance to aid their decision-making, sustained preparedness can be facilitated [40]. Such an initiative has been developed in Wales, for example, with Flood Awareness Officers working with residents on a one-to-one basis, advising on flood risks and assisting in the preparation of personal flood plans [41]. Similarly, a study in Germany employed a series of interactive workshops, bringing together the residents and the authorities around a discussion table; an observable shift from a blame discourse to an acceptance of responsibility by the community members was reported [42].

Although direct experience of flooding itself cannot, of course, be delivered to a community in order to influence perception, differing kinds of vicarious experience can be provided: for example, a project in the UK to engage local communities with historic flood events in their area, included oral histories, photographs and a presentation on a tsunami that occurred in the seventeenth century [43]. Another method, developed in Germany, makes use of a transparent water-tight box containing items of furniture to resemble a domestic lounge; a volunteer enters the room and water is introduced mimicking the flooding process, including the reactions of the volunteer as the depth increases [44]. This contributes to a more vivid visualisation of the flood experience by the spectators. By means of innovative approaches such as these, the 'past actions' components discussed earlier can be modified to some degree, by creating a personalised experience for those involved.

7 Conclusion

As alluded to earlier, the Breakwell analysis framework comprises a number of other factor-groupings, ten in all [9]. Not only are these too numerous to examine in detail in the present context but, more importantly, they are too numerous to include in any single experimental design. For pragmatic reasons, it is suggested that investigations into correlations between factors should focus upon a limited number of variables, whilst acknowledging that others exist [9]. Many models already exist in the field of risk behaviours, but as these incorporate factor groups from different sections of the framework, any attempt to superimpose them, in order to achieve an integrated model, would probably result in something resembling an inordinately complicated Venn diagram. It could, therefore, reasonably be argued that no single model is 'better' than any other where disparate groupings apply; any expectation that a 'correct' model of human behaviour in the face of risk will be forthcoming may, therefore, be over-optimistic.

For the time being, therefore, the issue of primary importance to policy-makers, namely to enable those at flood-risk to make appropriate preparations before flooding occurs, is being addressed: community engagement programmes, particularly those incorporating vicarious experience techniques, are able to bring about improvements in the preparedness behaviours of individuals, and thus increase the resilience of communities. Discussions regarding the scientific underpinnings of the process may continue indefinitely, but science has provided at least one means by which lives and property are better protected.

References

[1] Slovic, P., B. Fischhoff, and S. Lichtenstein, Perceived risk: psychological factors and social implications. *Proceedings of the Royal Society of London: Series A, Mathematical and Physical Sciences,* **376**: p. 17-34, 1981.

[2] Defra, *Changing behaviour through policy making*. Defra: London, 2005. Available from: http://www.defra.gov.uk/sustainable/government/documents/change-behaviour-model.pdf.

[3] Parker, D.J., S.J. Priest, and S.M. Tapsell, Understanding and enhancing the public's behavioural response to flood warning information. *Meteorological Applications,* **16**: p. 103-114, 2009.
[4] Fischhoff, B., Risk Perception and Communication Unplugged: Twenty Years of Process. *Risk Analysis,* **15**(2): p.137-145, 1995.
[5] Lave, T.R. and L.B. Lave, Public perception of the Risks of Floods: Implications for Communication. *Risk Analysis*, **11**(2): p. 255-267, 1991.
[6] Oldershaw, C. *The Earth in our hands: how geoscientists protect the public.* 2001. Available from: http://www.geolsoc.org.uk/webdav/site/GSL/shared/pdfs/education%20and%20careers/floods.pdf.
[7] Environment Agency. *River Severn - regulation fact sheet 2*. (nd) Available from: http://www.environment-agency.gov.uk/static/documents/Leisure/MIDS_Severn_Fact_sheet_2.pdf.
[8] Breakwell, P.G.M., *The Psychology of Risk.* Cambridge: Cambridge University Press, 2007.
[9] Harries, T., *Householder responses to flood risk; the consequences of the search for ontological security.* Middlesex University: London, p. 255, 2007.
[10] Brilly, M. and M. Polic, Public perception of flood risks, flood forecasting and mitigation. *Natural Hazards and Earth System Sciences*, **5**: p. 345-355, 2005.
[11] Norwich Union, *Homeowners fear future flooding but fail to take measures to protect their properties.* 2008. Available from: http://www.norwichunion.com/press/stories/3981-homeowners-fear-future-flooding-but-fail-to-take-measures-to-protect-their-properties.htm.
[12] Lorenzoni, I., S. Nicholson-Cole, and L. Whitmarsh, Barriers perceived to engaging with climate change among the UK public and their policy implications. *Global Environmental Change,* **17**(3-4): p. 445-459, 2007.
[13] Dupuis, A. and D.C. Thorns, Home, Home Ownership and the Search for Ontological Security. *The Sociological Review*, **46**(1): p. 24-47, 1998.
[14] Harries, T., Feeling secure or being secure? Why it can seem better not to protect yourself against a natural hazard. *Health, Risk & Society,* **10**(5): p. 479-490, 2008.
[15] Slovic, P. and E.U. Weber, *Perception of Risk Posed by Extreme Events, in Risk Management strategies in an Uncertain World.* Palisades, New York, 2002.
[16] Slovic, P., B. Fischhoff, and S. Lichtenstein, *Why study risk perception?* Risk Analysis, 1982. **2**(2): p. 83-93.
[17] Slovic, P., Perception of Risk. *Science*, 1987. **236**: p. 280-285.
[18] Slovic, P., *The Perception of Risk.* Risk, Society and Policy, ed. R. Lofstedt, London: Earthscan, p. 473, 2007.
[19] Sjöberg, L. *Risk perception is not what it seems: The psychometric paradigm revisited*, in VALDOR Conference 11 March 2003. Stockholm: VALDOR, 2003.
[20] Siegrist, M., T.C. Earle, and H. Gutscher, eds. *Trust in Risk Management: Uncertainty and Scepticism in the Public Mind.* Earthscan: London, 2007.

[21] Tompkins, E.L., L.-A. Hurlston, and W. Poortinga, Disaster Resilience: Fear, Friends and Foreignness as Determinants of Risk Mitigating Behaviour in Small Islands. in *SRI PAPERS,* Sustainability Research Institute, School of Earth and Environment, University of Leeds, 2009.
[22] Terpstra, T., M.K. Lindell, and J.M. Gutteling, Does communicating (flood) risk affect (flood) risk perceptions? Results of a quasi-experimental study. *Risk Analysis*, **29**(8): p. 1141-1155, 2009.
[23] Weinstein, N.D., Effects of Personal Experience on Self-Protective Behavior. *Psychological Bulletin*, **105**(1): p. 31-50, 1989.
[24] Paton, D., Houghton, B.F., Gregg, C.E., Gill, D.A., Ritchie, L.A., McIvor, D., Larin, P., Meinhold, S., Horan, J. and Johnston, D.M. Managing tsunami risk in coastal communities: identifying predictors of preparedness. *The Australian Journal of Emergency Management*, **23**(1): p. 4-9, 2008.
[25] Adeola, F.O., Katrina Cataclysm - Does Duration of Residency and Prior Experience Affect Impacts, Evacuation, and Adaptation Behavior Among Survivors? *Environment and Behavior*, **41**(4): p. 459 – 489, 2009.
[26] Horney, J.A., Snider, C., Malone, S., Gammons, L. and Ramsey, S. Factors Associated with Hurricane Preparedness: Results of a Pre-Hurricane Assessment. *Journal of Disaster Research*, **3** (2): p. 1-7, 2008.
[27] Sattler, D.N., C.F. Kaiser, and J.B. Hittner, Disaster Preparedness: Relationships Among Prior Experience, Personal Characteristics, and Distress. *Journal of Applied Social Psychology*, **30**(7): p. 1396-1420, 2000.
[28] Spittal, M.J., McClure, P.J., Siegert, R.J. and Walkey, F.H. Predictors of two types of earthquake preparation - Survival activities and Mitigation activities. *Environment and Behaviour,* **40**: p. 798-817, 2008.
[29] Lindell, M.K. and C. Prater, Risk area residents' perceptions and adoption of seismic hazard adjustments. *Applied Social Psychology*, **32**: p. 2377-2392, 2002.
[30] Knocke, E.T. and K.N. Kolivras, Flash Flood Awareness in Southwest Virginia. *Risk Analysis,* **27**(1): p. 155-169, 2007.
[31] Bandura, A., Self-efficacy: Toward a unifying theory of behavioral change. *Psychological Review*, **84**(2): p. 191-215, 1977.
[32] Ajzen, I., Perceived behavioral control, self-efficacy, locus of control and the theory of planned behaviour. *J. of Applied Social Psychology*, **32**(4): p. 665-683, 2002.
[33] Paton, D., Disaster preparedness: a social-cognitive perspective. *Disaster Prevention and Management,* **12**(3): p. 210-216, 2003.
[34] Harries, T., *Review of the Pilot Flood Protection Grant Scheme in a Recently Flooded Area (R&D Technical Report FD2651/TR),* in Delivering benefits through evidence, Department for Environment Food and Rural Affairs (Flood Management Division), Department for Environment, Food and Rural Affairs: London, 2009.
[35] Huber, M., *Reforming the UK Flood Insurance Regime - The Breakdown of a Gentlemen's Agreement.* ESRC Centre for Analysis of Risk and Regulation: London. p. 6, 2004.

[36] National Flood Forum, *Insurance industry flouting agreement with government*. Newsletter – National Flood Forum, 9, 2009.
[37] Speller, G., *Improving community and citizen engagement in flood risk management decision making, delivery and flood response - R&D Technical Report SC040033/SR3*, Environment Agency: Bristol. 2005.
[38] Paton, D. and D. Johnston, eds. *Disaster resilience - an integrated approach*. Charles C Thomas: Springfield, USA, 2006.
[39] Mc Ivor, D., D. Paton, and D. Johnston, Modelling Community Preparation for Natural Hazards: Understanding Hazard Cognitions. *Journal of Pacific Rim Psychology,* **3**(2): p. 39-46, 2009.
[40] Environment Agency (Wales). *Flood Awareness Wales*. (nd) Available from: http://www.environment-agency.gov.uk/113810.aspx.
[41] Pasche, E., Lawson, N., Douglas, I., Ashley, R., Garvin, S. and Schertzer, D. *Flood Risk Management in Small Urban Catchments.* in ERA-NET CRUE Funding Initiative: Mid-term seminar. Lyon, 17 October 2007.
[42] McEwen, L., *Community engagement with its flood history: understanding risk (Lower Severn Community Flood Information Network Final Report to the Royal Society – October 2006)*. Royal Society: London, 2006.
[43] Manojlovic, N. *Hamburg flood animation box.* (nd) Available from: http://daad.wb.tu-harburg.de/?id=1296.

A framework for participatory assessment of vulnerability of commercial property values to flooding in the UK

N. Bhattacharya[1], J. Lamond[2], D. G. Proverbs[2],
F. Hammond[1] & D. Searle[1]
[1]*School of Technology, University of Wolverhampton, UK*
[2]*Faculty of Environment and Technology, University of West of England, UK*

Abstract

The characteristics of flood risk situation in the UK for commercial properties involve large scale complexity in terms of effective contributory variables, different levels of spatio-temporal scales, and affected stakeholders' involvement. In the process of risk assessment, vulnerability has emerged as one of the most critical concepts as an indicative factor of measurement of severity of the likely consequence. The study builds on evaluation of existing approaches of participatory vulnerability and valuation approaches and moves on to develop new pluralistic participatory vulnerability analysis (PVA) method for analysis of qualitative and quantitative information. The focus is to investigate the effectiveness of participatory vulnerability analysis (PVA) approach supported by GIS as a potential tool for the UK commercial property sector for assessment of vulnerability of property values towards flooding. This includes the involvement of affected stakeholders to evaluate the predictive effect of the contributory factors on vulnerability of commercial property values. The proposed participatory spatial analysis method is suitable to be implemented within a Geographic Information System (GIS) framework. The approach is simple and transparent allowing public participation as well as accommodating data gaps by including opinions and experiential knowledge in the analysis.
Keywords: commercial properties, GIS, PVA, spatial analysis, valuation.

1 Introduction

In England and Wales approximately 185,000 commercial properties are at risk of flooding and average damage attributed to inland flooding is approximately £1Billion annually. Bennett [1] reported that this spending could exceed up to £27 Billion by 2080 .

As units of study in disaster research, commercial properties have recently made their place. Much of the research in the field is concentrated on residential properties and few have included empirical research of the commercial property sector [2]. Various traditional and modern methods of valuation are available which greatly depend on availability of market data. Each of these methods has their own advantages and disadvantages in relation to data availability and condition of usage. Pagourtzi *et al.* [3] and Shapiro *et al.* [4] noted that valuation studies of commercial properties in the UK have relied on knowledge and judgement of valuers. This is mainly because of difficulty in accessibility of valuation data at public domain and diverse and complex nature of data. Lack of data has affected valuation techniques and valuers often adopt options like available alternative comparable property data evidences [4, 5]. Spatial analysis method is one of the advanced methods which have a large potential in predictive modelling especially where data for comparative analysis of properties are sparsely available which the case in the UK.

Vulnerability is a dynamic concept which exposes the intrinsic feature of the community indicating a damage potential with multitude of components and forward looking variables (UNU-EHS personal communication, 2004 cited in Thywissen [6]). Vulnerability assessment for flood affected populations and properties are common in disaster research but are rarely used for commercial property valuation studies. It is highly relevant to have a comprehensive methodology to overview and analyse the potential impact flood vulnerability has on the value of commercial properties [7–9]. The motivation is to identify robust techniques that are sustainable to the changing physical and economic environment and reflect the market condition.

The term "vulnerability" of property is complex and involves a number of physical and social factors. Kenyon [10] reported that there are uncertainties involved in the assessment of vulnerability related to social perspective, where various stakeholders have conflicting interests, multiple legitimate perspectives, diffused responsibilities and differing level of impacts . Arguments have been put forward that under such complex circumstances, with variable criteria, a participatory approach is crucial for proper assessment and management [11, 12]. There is also increased emphasis on the participatory role of local government and stakeholders in flood risk management. However the challenge is to find the appropriate method or combination of methods where participatory approaches can be utilized and can be most effective for specific objectives. The main focus of the research is to analyse the effectiveness of the application of participatory vulnerability analysis (PVA) process to UK commercial property sector as a potential tool for assessment of vulnerability of property values towards flooding.

The paper is structured in three sections. First, description of the existing method of vulnerability assessment and Participatory Vulnerability Analysis is done. An evaluation of the methods is performed in the next section to recognize their suitability to the set objectives and identify potential enhancements. Based on this rationale the next section describes a novel method and the features of this new method that address issues identified associated with existing method. Finally, recommendation for implementation of the proposed method for vulnerability assessment of commercial properties in the UK is discussed before concluding.

2 Description of existing methods

2.1 The concept of vulnerability assessment

In the process of risk assessment, vulnerability has emerged as one of the most critical concepts as an indicative factor of measurement of severity of the likely consequence [13–15]. Vulnerability assessment has the advantage of involving multiple components to provide a holistic view of the level of risk and its value can be communicated among the stakeholders easily [16]. However, the number of indicators required for vulnerability assessment is highly data demanding and intensive in nature which often poses problem for the researchers. It is recognized that some of the criteria are either redundant or have no significant influence on the result.

Balica and Wright [16, 17] accounted that the number of indicators can be reduced to a manageable level based on the spatial scale of flood occurrence. Flood vulnerability assessment is useful if it is based on the most significant indicators with the focus on obtaining knowledge from the participation of stakeholders and allowing the end users to simplify their work as far as possible.

The actual factors are very site specific, location and hazard dependent [18]. The standard practice in this regard is to use the deductive approach, which is to follow a conceptual framework; this involves understanding the phenomenon followed by identification and selection of vulnerability indicators [17, 19, 20].

This criterion of selection of preliminary suitable indicators has been adopted for the proposed methodology. The selection of criteria and their suitability to local conditions can be backed up by outcome of expert survey. As a result incorporation of participatory approaches in this process is crucial.

2.2 Participatory approaches

In terms of complex environmental problems such as flood risk management, where several stakeholders and large number of properties with high potential of damage are involved, participatory approaches are increasingly recommended [21–23].

Actionaid [18] defined the concept of Participatory vulnerability assessment (PVA) as a systematic approach with the involvement of stakeholders in a thorough analysis of their level of vulnerability to empower and motivate them for appropriate future actions of preparedness and response to long term development . Since the nature of vulnerability is complex, and it requires predictive judgement, the analysis needs detailed and to the point breakdown. This is possible by dividing the process into well-defined phases as shown in figure 1.

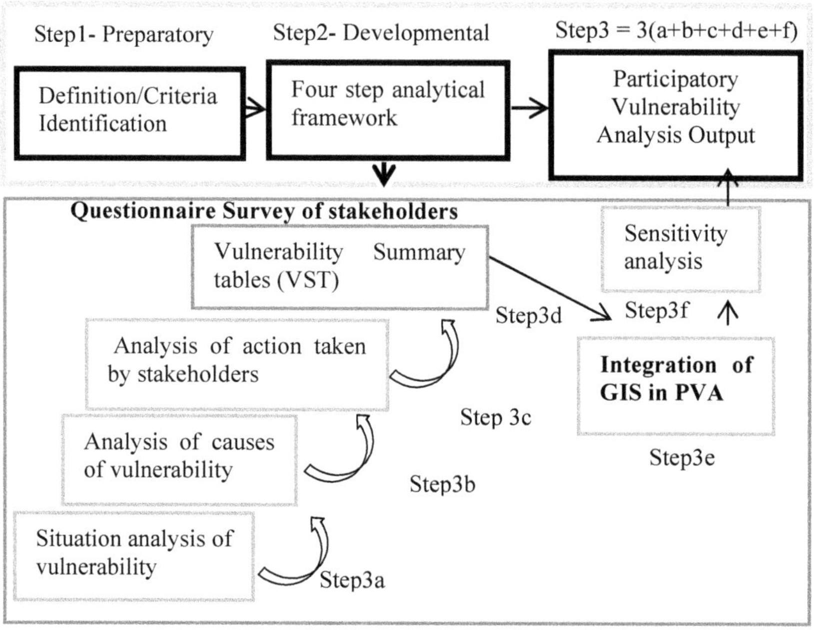

Figure 1: Participatory vulnerability analysis phases.

The stakeholders are given chance to weight the factors which are defined within the set questionnaire, according to their level of knowledge and experience. At the end of the participatory stage a complete matrix (vulnerability summary table or VST) of impact category, likelihood, impact details along with their weights should ideally be achieved to continue to the next stage. This phase will act as a base for the stages to follow where all these related information can be included in the generation of final composite map of vulnerability using Geographic Information System (GIS). Therefore, a spatial analysis technique supported by Participatory vulnerability analysis approach is proposed here as an appropriate methodology to assess the impact of flood vulnerability to commercial property value in the UK.

3 Evaluation of the proposed approach

The proposed new approach is designed to work with the local stakeholders involved in commercial property sector to conduct vulnerability assessment and identify the major drivers that affect property value in case of flood vulnerability.

Research has confirmed that there is a need to embed participatory processes in vulnerability assessment to benefit fully from the process and its wider relevance in the UK commercial property sector. PVA approach has benefit of providing transparency to the participants in understanding the different contributory factors of vulnerability. Prabhu *et al.* [22] described the process... is considered to be "simple, intuitive and transparent" and has the ability to deal with both qualitative and quantitative data. It has further scope of including experiential knowledge within the database in case of non-availability of observational data; therefore PVA is selected as a particularly appropriate methodology especially for markets where data is scarce as is the case in the UK.

Concerns related to the reliance on the knowledge of stakeholders in weighting the factors affecting their level of vulnerability may arise. Appropriate weighting of different factors according to their level of impact is necessarily subjective because the relationships between different factors are not linear and there are number of uncertainties involved. However, Muller *et al.* [24] noted that incorporation of information from two different sources (literature and stakeholders) could be used to strengthen the evidence for judgements as to how the scores may be assigned. Sensitivity analysis of outputs could be the answer to understand differential impact of differently scored factors on the outputs. It becomes especially important to understand the transferability of results for different stakeholders' perception.

Spatial data was rarely used in previous researches in the field of vulnerability of property value especially in the UK. Some of the methods used previously had problems like narrow domain of selecting implicit number of comparable properties from immediate localities and lacked in exhibiting spatio-temporal knowledge representatively [25–27]. The spatial analysis method, on the other hand, has great analytical capacity with the use of spatial data and interpolation techniques. This approach provides potential for a 'value range' maps which can be utilized for comparable properties in the form of a surface determined by the best representative data using GIS.

GIS can be used where various complex factors are involved as a potentially important tool for analysing digital data in space and time. This is because GIS has the ability to combine information of various spatial nature and source in one platform and illustrate it in an easily understandable manner to the concerned stakeholders in the form of maps. Pagourtzi *et al.* [3] and Wyatt [5] emphasized the benefits of GIS in their research in generation of 'value maps' including spatial distribution of values along with other property attributes. They also stressed on the potential use of GIS interface in value factor selection, units of comparison and analysis.

Pagourtzi *et al.* [3] and Lamond [25] indicated that the UK has lack of statistically significant sample sizes, comparative sales data of properties, therefore technique like spatial analysis are useful in such case .The proposed approach has advantage over the other approaches as it allows to go beyond the traditional way of explaining impact on property value using a set of man-made and ecological variables but also incorporate spatial heterogeneity, frequent updating using updated digital data without the need for repeated survey and greater analytical potential in research using GIS. Wyatt [27] pointed out that when valuation attributes are combined with flood vulnerability maps incorporating participatory approaches they can prove to be powerful decision making tools . They are also useful for purposes like planning, taxation, asset management, zonation of high and low value zones of properties (with particular reference to flood risk), recording changes in valuation over time and in assisting future valuation [5, 27].

4 Method design

4.1 Participatory Vulnerability Analysis (PVA)

The method is partly designed based on the framework guided by Actionaid International [18]. The purpose of this research focuses on analysis of the level of vulnerability and their root causes within a certain vulnerable group i.e., the commercial property sector. The methodological approach contains a flow of criteria which is guided by problem identification with specified objectives, problem structuring through screening in the form of VST. The VSTs include information related to impact data, qualitative and quantitative in nature. Based on these VSTs, type and category of impacts are identified for different levels of vulnerability of properties. The impact types can be categorized into physical, economic and social; while the impact categories are not fixed they differ based on distinct nature of impact and characteristics of the area. A questionnaire formulation is necessary including all the factors and criterion within each category. Data for the VSTs are collected mainly based on the survey.

At this stage it is important to involve the concerned stakeholders for instance the property occupiers to gain information and knowledge which can be compared with the existing literature based on the work flow showed in figure 1. For this reason, it is useful to start with a pilot study or smaller group. This can help in ensuring that the design for assessment has taken into account the most important factors. This also provides an option for screening out those categories of factors which do not comply with the scope of the study.

The process of ranking is allocated within the questionnaire that enables the participants (stakeholders) to rank the criterion based on their own judgement and weights can be derived from this information at a later stage of analysis. The summary of the VST provides different options for analysis of the effect of level of flood vulnerability on properties. These options for analysis are important aspects for understanding the potential impact on valuation as they have a futuristic component which involves a predictive quality. The process of PVA provides the base for the next stage which is the integration of the gathered information to assign the level of vulnerability to a composite vulnerability map by integration of GIS.The novelty in the approach is that the method is used to assess vulnerability of property values based on their level of vulnerability identified through integration of participatory approaches.

4.2 Integration of GIS in the assessment of vulnerability analysis

The contributing factors or selection of evaluation criteria for flood impact are categorized into physical, economic, social, technological and resilient attributes. These factors are taken into account in the frame of contribution towards vulnerability of flooding. Each factor is represented by a layer to form criteria maps which make the factors flexible to changes and highly transparent for the users to notice their effect on property values. Subsequently attributes are added to the layers to describe the information related to the causal factors.

Information obtained from the participatory approach are weighted according to their importance and added as layers for each contributing factor towards vulnerability. Muller *et al.* [24] described that to make the various maps comparable standardization of the ranked maps is necessary .Valuation data for commercial properties on a temporal basis are then formatted to be used in GIS environment. This data helps to create 'value range' map utilizing other attributes to illustrate the geographical distribution of property values over space and time. The value data is also helpful in spatial value pattern analysis, and autocorrelation analysis of property values. Such analysis can be a useful in predictive vulnerability assessment for property values where data is sparse.

4.2.1 Sensitivity Analysis

Sensitivity Analysis (SA) is a prerequisite for any model since that determines the reliability of the model in assessment of uncertainties. In this case, the different input parameters are systematically varied based on their weights provided by different groups of stakeholders to understand the change in behaviour of the outputs.

Edgar [26] emphasized that pairwise comparisons between most relevant factors can be useful for carrying out such analysis . It decomposes the problem in a hierarchical manner with comparative judgement and synthesis of priorities.

Chen *et al.* [28] reported that the pairwise comparison method of the weighted layers will create a ratio matrix to view all the paired criteria and compare the different outputs, make changes to the input ranks and calculate consistency based on user's preferences . Sensitivity is determined by comparing dependency of the input parameters and the changed parameters to see if there is any significant impact on the final outputs.

5 Conclusion

Relatively little attention has been paid to how to conduct assessment of vulnerability of commercial property values to flooding in valuation or flood management research. In the light of changing flood risk and flood management policy in the UK it was found to be important to conduct such assessments in a way which not only reflects the current conditions but also helps to build capacity for the local communities by reflecting their views and opinions within the system.

The structured approach of the proposed variation to the PVA methodology provides a well-focussed agenda, in the form of complete breakdown of the required tasks of vulnerability assessment and incorporation of information for both qualitative and quantitative options to improve transparency in the process.

The proposed method of spatial analysis integrates existing approaches and provides a spatio-temporal dimension by incorporating GIS. The methodology is in its developmental stage and anticipated to perform well if the data and other relevant information can be made available for analysis. The collection of data in proper GIS format might add complexity to the given approach; however with increasing level of data in digitised format, the development of appropriate approaches exploiting the potential of such data is encouraged. It is expected that an integrated purpose-built pluralistic assessment of flood vulnerability of commercial properties towards property values can be successfully performed involving stakeholders views in evaluation of alternatives for decision making in this field in future.

References

[1] House of Commons library research. *Reducing Flood Risk*. London. 2011.

[2] Kenny, S., Pottinger, G., F. and Yasmin, P., *Flood risk and property,impacts on commercial & residential stakeholders' strategy*. College of Estate Management: Reading. 2006.

[3] Pagourtzi, E., Assimakopoulos, V., Hatzichristos, T., French, N., *Real estate Appraisal: A review of valuation Methods*. Journal of Property Investment and Finance, **21**(4): p. 383-401. 2003.

[4] Shapiro, E., Davies, K., Mackmin, D., *Modern Methods of Valuation*. 10 ed. Malta: Melita Press. 701. 2009.

[5] Wyatt, P., *Using a geographical information system for property valuation*. Journal of Property Valuation and Investment, **14**(1): p. 67-79. 1996.

[6] Thywissen, K., *Components of Risk, in Studies of the University: Research, Counsel, Education- Publication series of UNU-EHS*. UNU Institute for Environment and Human Security: Bohn. p. 52. 2006.

[7] ROOTS. *The property implications of flood risk management, RICS practice information and guidance. in RICS rural research conference*. Brighton. 2010.

[8] Environment Agency. *Flooding in England: A national assessment of flood risk*. Bristol. 2009

[9] UK Government. *Flood and Water Management Act*. London. 2010.

[10] Kenyon, W., *Evaluating flood risk management options in Scotland: A participant-led multi-criteria approach*. Ecological Economics: p. 70-81. 2007.

[11] Sybille, v.d.H., *Participatory approaches to environmental policy-making: the European Commission Climate Policy Process as a case study*. Ecological Economics, **33**(3): p. 457-472. 2000.

[12] Environment Agency. *Participatory risk assessment: Involving lay audiences in environmental decisions on risk*. Reading. 2003.

[13] Bhattacharya, N., Lamond, J., Proverbs, D., *Flood vulnerability and hazard adjustment for UK commercial sector, in International Conference on Disaster Resilience:* Srilanka. 2011.

[14] Vatsa, K.S., *Risk, Vulnerability and asset based approach to disaster risk management* International Journal of Sociology and Social policy, **24**(10/11): p. 1-48. 2004.

[15] Zhang, Y., M.K. Lindell, and C.S. Prater, *Vulnerability of community businesses to environmental disasters.* Disasters, **33**(1): p. 38-57. 2009.

[16] Balica, S. and N.G. Wright, *A network of knowledge on applying an indicator-based methodology for minimizing flood vulnerability.* Hydrological Processes, **23**(20): p. 2983-2986. 2009.

[17] Balica, S., Wright, G. N., *Reducing the complexity of flood vulnerability index.* Environmental Hazards, **9**: p. 321-339. 2010.

[18] Actionaid, *Participatory Vulnerability Analysis - A step by step guide for field staff.* Actionaid International: London. 2005.

[19] Penning-Rowsell, E.C., Wilson, T, *Gauging the impact of natural hazards: the pattern and cost of emergency response during flood events.* Transactions of the Institute of British Geographers, **31**(2): p. 99-115. 2006.

[20] Messner, F., Meyer, V., *Flood damage, vulnerability and risk perception-challenges for flood damage research, in Discussion paper.* UFZ: Leipiz. 2005.

[21] Sigrid, S., *Multicriteria evaluation and public participation: the case of UK energy policy.* Land Use Policy, **23**(1): p. 53-62. 2006.

[22] Prabhu, R., Sukadri, D., Purnomo, H., Hartanto, H., Guidelines for applying multi-criteria analysis to the assessment of criteria and indicators. CFOR: Jakarta. 1999.

[23] Petts, J., Homan, J., Pollard, S., *Participatory Risk Assessment: Involving Lay Audiences in Environmental Decisions on Risk.* Environment Agency: Reading. 2003.

[24] Muller, A., Reiter, J., Weiland, U., *Assessment of urban vulnerability towards floods using an indicator-based approach – a case study for Santiago de Chile.* Natural Hazards Earth System Science, **11**: p. 2107-2011. 2011.

[25] Lamond, J.E., *The impact of flooding on the value of residential property in the UK, in Dept. of build environment and engineering.* University of Wolverhampton: Wolverhampton. p. 377. 2008.

[26] Edgar, M., *Expert systems in investment decision-making.* Estates Gazette. p. 62-64. 1994.

[27] Wyatt, P., *The development of a property information system for valuation using a geographical information system (GIS).* Journal of Property Research, **13**(4): p. 317-336. 1996.

[28] Chen, Y., Yu, J., Shahbaz, K., Xevi, E. *A GIS-Based Sensitivity Analysis of Multi-Criteria Weights in 18th World IMACS / MODSIM Congress.* Cairns, Australia. 2009.

Sustainable flood memories, informal knowledge and the development of community resilience to future flood risk

L. J. McEwen[1], F. Krause[2], O. Jones[1] & J. Garde Hansen[2]
[1]*University of the West of England, UK*
[2]*University of Gloucestershire, UK*

Abstract

The UK policy change from 'flood defence' to 'flood risk management' in the 1990s has involved shifts to more distributed flood risk management responsibilities. This poses questions about roles of floodplain residents in community-led adaptation planning for changing flood risk, and how these roles can be supported/strengthened. Research evidence emphasises importance of informal/local/lay knowledge, 'watery sense of place', flood memories and shared flood heritage, in how communities prepare for, and recover from, floods. This paper outlines initial research outcomes from an interdisciplinary UK Economic and Social Research Council-funded research project that proposes the concept of 'sustainable flood memory' in the context of effective flood risk management. This is conceived as an approach to memory work that is community focused, archival, integrating individual and collective experiences, involving inter- and intra-generational communication, and strategies for its future. The project aims to increase understanding of how flood memories provide a platform for developing informal knowledge, so creating social learning opportunities in communities that can increase their adaptive capacities and flood resilience. It explores: how communities remember and archive flood experiences; how local flood knowledge is materialised, assimilated and protected; the role of catastrophic floods (e.g. July 2007, River Severn, UK) in building 'community' memory and flood risk knowledge; and whether informal knowledge is tacit or otherwise, if and how it is learned, and whether it can be

WIT Transactions on Ecology and The Environment, Vol 159, © 2012 WIT Press
www.witpress.com, ISSN 1743-3541 (on-line)
doi:10.2495/FRIAR120211

transmitted, developed, revitalised and returned in settings where it is lost or lacking.
Keywords: adaptive capacity community engagement, flood risk, informal knowledge, resilience, sustainable flood memory.

1 Introduction

This paper outlines UK Economic and Social Research Council funded research (ESRC) that is investigating the importance of sustainable flood memory, with its associated 'watery sense of place', shared flood heritage, folk memories of flooding, local flood knowledge, community flood awareness and opportunities for social learning (see ESRC *Flood Memories* website/blog [1]). How do these different elements interact to influence the propensity for local action to increase community resilience to future flooding? The interdisciplinary research connects geography, environment, heritage, media, and cultural and memory studies. The research is evaluating the extent to which communities that have a history of past flood events are more resilient to future flood events than communities with no previous history of flooding, or no shared memories of flooding. This paper starts by setting the research contexts, proposes the concept of sustainable flood memory, outlines aims and methods, and identifies emerging themes from initial project results.

1.1 Recent UK flood events, policy response background and local communities

The policy change from 'flood defence' to 'flood risk management' (FRM) after the 1998 and 2000 floods (rivers Avon, Severn, UK; Tunstall *et al.* [2]; Johnson *et al.* [3]) involved a shift to more distributed responsibilities for UK flood risk management. This is set in context of climate change scenarios that predict increased risk of specific types of flooding (UKCIP [4]), and major questions over sustainability of hard engineering defence solutions. The catastrophic floods (Marsh and Hannaford [5]) and management crises in Hull and the Severn catchment in July 2007 proved to be one of the greatest civil emergencies that the UK has faced, with loss of life, houses/businesses devastated, and strategic transport, power and water infrastructures threatened, damaged and disabled. In the aftermath, a whole range of agencies and tiers of government have reviewed procedures and planning in FRM from local to national levels (Cabinet Office [6]; DEFRA [7]). Several imperatives (e.g. need for adaptive capabilities after floods; community lead adaptation planning) have emphasised research needs to establish how floodplain groups can be made less vulnerable, and more sustainable and resilient (Tobin [8]).Therefore, investigating the relationships between sustainable flood memory, local knowledge, and how individuals/communities deal with risk and effective community level planning in developing local resilience, is important, whilst recognising that the concept of 'community' is itself contested. This involves exploring new and different perspectives on the sociology of risk, drawing on different literatures around changing meanings of disasters (e.g. Furedi [9]), with crises as catalysts of change and cultural memory.

1.2 Distributed flood risk management

Distributed FRM involves all stakeholders taking some degree of responsibility for their own protection (e.g. by take-up of flood insurance, or flood resilience/resistance measures to their property; CIRIA [10]). If individuals are not aware of flood risk or how to respond in flood events then they will not prepare or deal effectively in emergencies. Even if they do perceive risk, research indicates that individuals may still not act rationally to protect themselves against future flooding (Baker [11]). Clark and Priest's [12] research on public awareness of flood risk, drawing on theories of 'reasoned action' and 'behaviour change', challenges notions that public information can alter public behaviour in clearly defined or predictable ways.

1.3 Local/lay/informal knowledge

It is increasingly recognised that local knowledge is articulated in various ways – as lay, traditional and indigenous knowledge (hereafter local knowledge), nested within notions of common wisdom, that have implications for mapping knowledge and the redistribution of expertise (Whatmore [13]; Lane *et al.* [14]). The 2007 floods challenged the role of formal FRM and 'expert' knowledge in how to manage fluvial floods (Cabinet Office [6]). One emergent element in the review process is that local flood knowledge and associated local senses/values of place are key elements in developing flood knowledge to build future community resilience (McEwen and Jones, [15]). Local knowledge at individual and community levels is increasingly considered important through the Disaster Response Spiral - in preparation and during floods, and in adaptive capabilities and post-flood learnings (Dufty [16]; McEwen [17]).

1.4 Community level planning

The increased focus on the 'local' is step-changing the roles of community groups in community-lead adaptation planning, and increasing the role of Local Resilience Forums. Local community action groups formed to campaign around flood risk issues on specific stretches of local water courses perceived themselves to be unheard by official FRM organisations prior to the 2007 events. The same groups are now being co-opted by expert agencies (e.g. water supply companies; environmental regulator) into planning and governance structures (McEwen and Jones [15]). Similarly, Parish Councils, and even concerned/ informed/ active individual citizens, are being supported and encouraged to become part of the resilience/planning structures. Their monitoring of water courses and drainage systems at the most detailed, local level provides local information that can link into wider 'expert' knowledge systems. This local knowledge is often intimately connected with memories of, and emotional responses to, past flood events which persist in a number of ways.

1.5 Loss of community knowledge

This incorporation of local communities and their knowledge into flood resilience planning needs to take into account the changing nature of UK communities (Coates [18]). These cannot be assumed to be homogenous or contiguous in rural or urban areas. Rather they are likely to be plural, multi-faceted, fluid and made up of groups with a disparity of resources. Vulnerable groupings (by age, socio-economic class, ethnicity, disability) may be particularly exposed to changing flood risk, and may also be those least likely to develop a sense of place and community memories of past flood events which help them deal with future flooding. There is anecdotal evidence that traditional community flood resilience has reduced or broken down in communities with transient or intermittent membership. This poses important questions as to how flood memories develop, what local flood knowledge comprises, and how it is developed and shared. This parallels how climate change is constructed differently by individuals and communities, in contrast to climatologists (Hulme *et al.* [19]).

2 The project

It is therefore urgent and timely to explore flood memories and how these provide the platform for developing character and role of local knowledge for personal and community resilience. This requires research that explores: (a) how communities remember and archive flood experiences; (b) how local flood knowledge is materialised, assimilated and protected; (c) the role of catastrophic floods in building community, community identity, and community knowledge of flood risk, in conjunction with more frequent, moderate floods; (d) whether informal knowledge is tacit or otherwise, if and how it is learned, with key factors in the social learning process; and (e) whether informal knowledge can be transmitted, developed, revitalised and returned in settings where for various reasons it is lost or lacking. The research is critically evaluating the extent to which, and in what ways, sustainable flood memories contribute to local flood knowledge, building community resilience in the face of changing flood risk and actual events. To achieve this, the research is comparatively studying three communities in the lower Severn valley, UK, which have different histories, forms and levels of past flood experience, and different degrees of community development.

2.1 Extreme floods and flood histories

The impact of the July 2007 floods in the UK highlighted the need for community capacity building for effective FRM. These floods were caused by sustained, high intensity rainfall over a wide area, combining fluvial, pluvial (rainfall) and groundwater flooding, with low annual percentage probabilities of occurrence (<0.5 in some catchments; Marsh and Hannaford [5]). The floods formed part of a 'flood rich' period on the middle to lower Severn, with severe floods in April 1998 and November/December 2000. In contrast, a 'flood poor' period occurred in 1970s and 1980s, with a proceeding flood cluster in the 1960s (McEwen [17]). The largest 20th century flood occurred in March 1947 - currently within living memory of some older floodplain residents.

2.2 Sustainable memories, senses of place and heritage

In the past few decades, memory studies have flourished and academic work has drawn upon the sociological and psychological approaches to personal (Bartlett [20]), collective (Halbwachs [21]), social (Fentress and Wickham [22]), and cultural (Sturken [23]) memory. This research proposes the concept of sustainable flood memory, which can be conceived as an approach to memory work that is community focused, archival, integrating individual/personal and collective/community experiences, involving inter- (vertical) and intra-generational (horizontal) communication and strategies for its future. The research aims to discover what factors link memory and local knowledge, and how these connect and disconnect during and after flood events. It explores how unravelling these factors can help understand processes of community or informal social learning that might be supported/enhanced in flood governance by agencies concerned with FRM at local level. These insights can then be used as a means to strengthen community resilience to future flood events in terms of actual flood impact and risk.

The evidence base for sustainable flood memory is taken to be the narratives, histories and folk memories of previous flood events and their impacts that are embedded in local communities and culture. This in turn, we suspect, can lead to what is a 'watery sense of place' - that is living with water and the risks it brings become part of collective understandings of place characteristics, distinctiveness and identity. While flooding is not exactly welcomed, living with floods and flood risk in terms of knowledge, expectation and resilience becomes part of peoples' individual, family and community identity. Some idea of this kind of understanding of place can found in Sutherland and Nicholson's [24] book *Wetland: Life in the Somerset Levels*, where living with flooding is depicted as a key marker of place identity.

3 Research methods

The lower Severn Valley was one of the worst affected areas in July 2007 floods; and where members of the research team already had extensive knowledge, previous research experience and contacts. The three communities studied comprise: (a) an 'established' community which has a significant history of episodic extreme floods, regular experience of flooding, and corresponding flood memories (Setting 1); (b) a 'newer' community which has no previous history of flooding (built after previous extreme floods) but which was flooded in July 2007 (Setting 2); and (c) one floodplain district with a past history of extreme flooding including recent experience in 2007, but with significant transient or intermittent occupation of residents, and thus less chance of developing long term community cultures/knowledge (Setting 3). In addition, a rural/ urban contrast was built into the research design.

The project is developing the model of a stakeholder competency group (SCG; cf. Langström, *et al.* [25]) as an integral part of the research process. Its principle is that key stakeholders can input to the research process, inform the analysis of results, and advise on dissemination processes to maximise research value and impact. The SCG comprises community flood action groups, the environmental regulator, local government, Parish and Community Councils (lowest tier of UK government), rural community development councils, National Flood Forum (national self-help group for those affected by flooding), insurance industry, and local archives/museums. Thirty residents are being interviewed in each case-study area, using a snowballing technique to identify interviewees. Several themes were considered in the semi-structured interviews:

- Background of interviewee
- 2007 floods: What happened?
- 2007 floods: How did you react?
- Previous flooding
- Memory and resilience
- Recording the flood
- Sharing stories, images and other records
- Remembering and forgetting
- Maintaining/developing memories.

4 Emerging research themes

Initial analysis of the interviews indicates that a large volume of personal flood memories and interest in floods exist in the three settings. Memories of floods cover a spectrum of experiences including revitalised community spirit, and diverse emotional responses, with awe, fear, excitement, grief and challenge all represented. The following synthesis briefly explores some emergent themes, with illustrative quotes.

4.1 Memory as process

Memory can be conceived as a vertical process as much as a defined body of knowledge experienced horizontally. This is reflected widely in the memory studies literature, from Halbwach's [21] classic analyses to current studies (e.g. Erll and Nünning [26]; Olick and Robbins [27]). A key part of that process is encouraging people to remember and pass down their knowledge, which they may already do in private settings. Creating opportunities for communities to remember helps develop memories; critical here is establishing the sites and frames of sustainable flood memory.

She: *'Some new people had moved into the street. We said, it would be nice for the new people to get to know the people who already lived here.'*

Interviewer: *'And in these occasions, do floods come on as a subject?'*

She: '*Yeah, sometimes.*'

He: *'Of course it does. You talk about it, you know. And then, that's the time when the bad things come out as well as the good things.'* (female, 63 years, and male, 62 years, Setting 3).

When the memory is experienced horizontally as a body of knowledge in a public setting, tensions can exist in promoting local flood memory in well-established communities.

'We have a church flower festival here every two years. [...] Our friend next door, he's the Chairman of the Flood Prevention Committee, and [...] he produced [...] photographs on big boards and wanted to put them in the churchyard [...] when the last flower festival was on, but he was turned down by the committee [...] he was allowed a small portion in a tent but he wanted to let everybody see, all the visitors, because they get up to 3,000 to 4,000 because it's over 3 days, [...] he wanted everybody to see what the situation had been like..'. (male; 73 years, Setting 1).

4.2 Diverse ways of memorialising/ materialising floods

Methods of materialising, memorialising and visualising floods vary widely. Flood marking of maximum 2007 flood levels against physical or known points of reference has occurred at several locations that have community resonance - on churches, public houses, teashops, garages. Flood materialisation occurs in newspaper cuttings; photography/film making; oral history/storytelling; and the emerging use of new technologies including social networking sites. Such materialisation varies in the extent to which it is official/unofficial, verifiable, and at personal or community levels. Physical locations for flood materialisation can be in individual houses, and collated in official and informal archives. Flood materialisation can occur in the everyday – through the resistance/resilience measures implemented post-flood in the home.

'We had a look outside her front door, which is always blocked with a flood barrier (that she is able to step over, though. Nevertheless, she seems to usually use her back entrance).' (interviewer notes – woman, 82 years, Setting 3).

4.3 Childhood memories of 'learning to live with floods'

Some interviewees, with longer residency, in remembering the 2007 floods, referred to childhood memories of living through floods in 1947 and the 1960s, and 'awareness/knowing what you can and cannot do' in flood water. These memories embody changing attitudes to living with water, and willingness to take risk. There was a strong view that people need to be aware of the possibility of flooding, and its dangers as part of a 'watery sense of place'.

'I've always been brought up around flooding. As a child, we used to walk the floods. We've done it on Christmas Day, to walk to the pub at Lower Lode across the embankments, the flood barriers. There was water rushing; it was quite deep. All holding hands linked, and mum had wrapped scarves around our hands. We were walking along and [name] was tied to the dog, my little sister. [...] He's a big dog, a big lurcher cross greyhound, so he's a big powerful dog.

[...] And then we all went to the pub for a drink. Well, they did; we all just played in the floods.' (male, 34 years, Setting 1)

'And then way back, ever since I was a youngster I can remember the floods coming in. And it was a regular thing that in February you always looked out to see, you'd look across the meadow as we called it. It hadn't been filled in [with the landfill site] then. And because the river would overflow and you keep your eye on it, you see oh it's coming up, coming up, get the furniture up, get the carpets up.' (female, 74 years, Setting 3).

Such childhood memories demonstrate a watery sense of place, with insights into past coping strategies and common wisdom that can be brought to present and future flooding. Opportunities exist for inter-generational learning around resilience strategies in past communities.

4.4 Memories of family/community response

Flood memories also revolved around the impact of floods on family, along with their reactions and responses to the situation and loss, and during flooding, the support from neighbours and friends.

*'When it was about three o'clock in the morning and there was left myself, my mum and dad, [name] who works for us, and he'd stayed to help, [name] who works for us, and he'd stayed to help... All the customers had gone. [name]'s wife was sat upon the bar. [...] Dad was behind the bar up to his waist in water. We all had drinks in our hands, and dad went, "that's it, **** to it. What's happened has happened." And then [name] started singing 'always look on the bright side of life' by Monty Python, while we were covered up to our waist! And we all just stood there – it's on video somewhere, just wish I could find it. [...]That will be something that I'll remember for the rest of my life'.* (male, 34 years, Setting 1)

4.5 How to develop flood memories

A key area of research exploration is how to develop flood memories in settings where they are lacking or lost. Some respondents suggested the importance of bringing up 'living with flooding' as an issue regularly, independently of actual flood events, e.g. through schools in the curriculum (e.g. a flood week); festivals linked more broadly to water heritage ("Water Festival"); exhibitions (at houses; local museums/ heritage centres, or in public spaces), and presentations/ fora that encourage community groups to share images, stories, videos.

'I said to them a little while ago, I had lots of pictures and would anybody be interested, and they said yes. And of course this is a building where people retire, so we have lots of people moving in and out. We have had lots of newcomers who wouldn't have a clue, so of course they were very keen to come and have a look, and they were astonished'. (female, 75 years, Setting 3)

Other proposals included the use of sculpture to remember washland areas as a key element of community identity. Some of these suggested community engagement methods have been used elsewhere to promote 'watery sense of place' (e.g. Shrewsbury's Water Festival).

4.6 Active remembering

Active remembering can occur through 'associations' or nuclei of people (e.g. community flood action groups) brought together during the 2007 floods (Hemming [28]). Flood Action Groups are frequently initiated and mobilised after extreme floods (e.g. Severn and Avon Combined Flood Action Group [29]). Rehearsal and reinforcement of flood memory can occur through their on-going campaigning activities to secure official inputs to their flood protection. This keeps flood memory high in the consciousness of both the association and other community members.

4.7 Active forgetting

Some people prefer not to remember the floods, because they associate them with painful/traumatic memories. Local knowledge can lack tuning (e.g. that intense rainfall always leads to extreme floods).

'I think you've got to actually try and forget them cause they were terrifying. And if you try... If you think too much about it then you... I mean, obviously for two or three years after those floods every bit of rain, every bit of flooding terrified some people, absolutely terrified them. They thought that this was all going to happen again.' (female, 76 years, Setting 2)

Others have 'moved on', with the 2007 flooding perceived as 'old news', unwilling to discuss their experiences. 'Active forgetting' poses questions about the characteristics of people who do not want to remember, and the implications for their adaptive capabilities and post-flood learnings, particularly when they are key players in communities.

What messages are conveyed with new or upgraded flood alleviation works? 'It will never happen again' or 'this is a flood risk area?' An engineering intervention can only be effective up to design limits – the well known 'levee effect' (Tobin [30]) – but can lead to the belief that residual risk is removed.

'And then after that in the 1970s we put up the new defences and it was working extremely well. People got more confident and the Council said okay, and some of the semi-derelict houses were bought and completely rebuilt and that sort of thing. So we were quite confident really.' (male, 75 years, Setting 1).

4.8 Flood memories as catalysts for action

Flood memory can act as a catalyst for individual and community action, through the implementation of individual resistance/ resilience measures or through local campaigning for 'official' assistance. For example:

'In my day, 40 years ago, we wanted to improve the flood banks then, and we had a sub-committee within the village that increased the height of the banks, with completely our own efforts really. [...] We were already existing with a pretty good set-up really, and then after the 2007 flood it was realised by the village that the flood defences wanted making much higher and much stronger and so this sub-committee was formed of five of us I think, and we then started making plans to get grants.' (male, 75 years, Setting 1).

4.9 Relationships with other stakeholders in FRM

In some urban floodplain groups, memories of the 2007 floods were articulated in terms of widespread disillusionment with, and distrust of, formal bodies (environmental regulator, local government). In others, flood memories embodied a more positive experience of working with official organisations to mitigate risk. At the same time, institutional memory in organisations with a stake in FRM can decay rapidly, with restructuring, staff changes/losses, and functions outsourced. The changing membership of our SCG reflects such personnel changes during the ESRC project.

5 Conclusions

The emerging evidence poses questions for resilience planning and increasing the adaptive capabilities and post-flood learnings of floodplain groups. Top-down flood risk management can assume and enhance the dependency of groups and individuals living on floodplains; distributed FRM requires empowered and well-informed communities. Our research has uncovered considerable local learning resources and data that support the concept of sustainable flood memory. This can be practiced vertically (between generations and over time) as much as horizontally (in the moment of flooding). It is the former memory process that is worth pursuing publicly.

Creating strategies for vertical community memories of flooding has potential to produce social learning between groups and over time, so informing approaches to community flood education. Horizontal community memories show up the diverse ways that different individuals materialise flood memory during a flood event in a discreet moment in time (oral recordings, artwork, videos, photographs, social networking, diaries, news reports). It is the connection of the horizontal and vertical axes to provide opportunities for inter-generational learning of past coping strategies that could be developed as a means of increasing community resilience in a flood risk context.

Active forgetting is only possible if community memory remains on the horizontal axis and memories are only passed down privately. Likewise, institutional memories need to understand how community memory functions and how best to engage local knowledge to increase the adaptive capacity/ post-flood learnings of floodplain groups. Contrasts can be made between the character and longevity of individual/ community and institutional memory, with implications for how organisations work with community to develop flood memory.

The next stages of the project combine participant observation at meetings, in depth focus groups with community, and interviews with key institutional stakeholders. Outcomes from the project will include two action packs for community and organisational stakeholders, co-produced with the stakeholder competency group, outlining how sustainable flood memory can be enhanced and developed as a means of increasing community resilience.

Acknowledgement

This research is funded by the UK Economic and Social Research Council.

References

[1] ESRC Sustainable flood memories. Available at: www.glos.ac.uk/flood memories; floodmemories.wordpress.com [accessed on 24th February 2012].
[2] Tunstall, S.M., Johnson, C.L. and Penning-Rowsell, E.C. *Flood hazard management in England and Wales: from land drainage to flood risk management*, Proceedings - World Congress on Natural Disaster Mitigation. The Institution of Engineers (India) - WFEO, India 2, 447-454, 2004.
[3] Johnson, C. L., Tunstall, S. M., and Penning-Rowsell, E. C. Floods as catalysts for policy change: historical lessons from England and Wales. *International Journal of Water Resources Development*, 21, 561-575, 2005.
[4] United Kingdom Climatic Impacts Programme (UKCIP). Available at www.ukcip.gov.uk [accessed on 1st May 2009].
[5] Marsh, T. J. and Hannaford, J. *The summer 2007 floods in England and Wales – a hydrological appraisal.* Centre for Ecology and Hydrology, Wallingford, UK, 2007.
[6] Cabinet Office *The Pitt Review: Lessons learned from the 2007 floods* http://www.cabinetoffice.gov.uk/thepittreview.aspx, 2008.
[7] DEFRA Government Response to Pitt Review (Dec 2008).
[8] Tobin, G.A. Sustainability and Community Resilience: The Holy Grail of Hazards Planning? *Environmental Hazards: Human and Policy Dimensions, Global Environmental Change*, Part B. 1, 13–25, 1999.
[9] Furedi, F. The changing meaning of disaster. *Area* 39.4, 482-489, 2007.
[10] Construction Industry Research and Information Association (CIRIA) Available at: http://www.ciria.org [accessed on 12th December 2009]
[11] Baker, V. Flood hazard science, policy, and values: A pragmatist stance. *Technology in Society* 29, 161–168, 2007.
[12] Clark, M. J. and Priest, S. J. *Public awareness of flood risk: the role of the Environment Agency flood map.* ESRC End of Award Report, RES-000-22-1710. Swindon: ESRC, 2008.
[13] Whatmore, S. J. Mapping knowledge controversies: science, democracy and the redistribution of expertise. *Progress in Human Geography* 33, 587–598, 2009.
[14] Lane S. N., Odoni N., Landström Whatmore S. J., Ward N. and Bradley S. Doing flood risk science differently: an experiment in radical scientific method. *Transactions of the Institute of British Geographers* 36, 15-36, 2011.

[15] McEwen, L. J. and Jones, O. Building local/lay flood knowledge into community flood resilience planning after the July 2007 floods, Gloucestershire, UK. *Hydrology Research, Special Issue* (in press).
[16] Dufty, N. A new approach to flood education. *The Australian Journal of Emergency Management* 23, 4-8, 2008.
[17] McEwen, L. J. Approaches to Community Flood Science Engagement: the Lower River Severn catchment, UK as Case-study. *International Journal of Science in Society* 2, 159-179, 2011.
[18] Coates, T. The Conscious Community: belonging, identities and networks in local communities' flood response. Unpublished PhD, Middlesex University, UK, 2010.
[19] Hulme, M., Dessai, S., Lorenzoni, I. and Nelson, D. R. Unstable climates: exploring the statistical and social constructions of 'normal' climate Geoforum 40, 197-206, 2009.
[20] Bartlett, F.C. *Remembering: An Experimental and Social Study*, Cambridge University Press, Cambridge, 1932.
[21] Halbwachs, M. *On Collective Memory*. Translated and edited by Lewis A. Coser. University of Chicago Press, Chicago, [1925] 1992.
[22] Fentress, J. and Wickham, C. *Social Memory* Blackwell, Oxford, 1992.
[23] Sturken, M. Tangled Memories: The Vietnam War, the Aids Epidemic, and the Politics of Remembering. Berkeley: University of California Press, 1997.
[24] Sutherland, P. and Nicholson, A. *Wetland: Life in the Somerset Levels*, Mermaid Books, Williamsburg, Virginia, 1986.
[25] Langström, C., Whatmore, S. J., Lane, S. N., Odoni, N. A., Ward, N. and Bradley, S. Co-producing flood risk knowledge: redistributing expertise in critical 'participatory modelling'. Environment and Planning A 43, 1617-1633, 2011.
[26] Erll, A., Nünning, A. (eds.), and S. Young, *A Companion to Cultural Memory Studies*. Germany: De Gruyter, Germany, 2008.
[27] Olick, J. K. and Robins, J. Social Memory Studies: From "Collective Memory" to the Historical Sociology of Mnemonic Practices. *Annual Review of Sociology* 24, 105-140, 1998.
[28] Hemming, H. Together: how small groups achieve big things. John Murray, London, 2011.
[29] Severn and Avon Combined Flood Group Building on the floodplain is misguided (a definitive proof of evidence). Unpublished report, 2008.
[30] Tobin, G. A. The levee love affair: a stormy relationship. *JAWRA Journal of American Water Resources* 21, 359-367, 1995.

Author Index

www.ingramcontent.com/pod-product-compliance
Ingram Content Group UK Ltd.
Pitfield, Milton Keynes, MK11 3LW, UK
UKHW020248290926
15470UKWH00002B/809

9 781845 645885